Reactivity and Structure
Concepts in Organic Chemistry

Volume 24

Editors:

Klaus Hafner   Jean-Marie Lehn
Charles W. Rees   P. von Ragué Schleyer
Barry M. Trost   Rudolf Zahradník

Manfred T. Reetz

Organotitanium Reagents in Organic Synthesis

With 23 Figures

Springer-Verlag
Berlin Heidelberg New York Tokyo

Professor Dr. Manfred T. Reetz

Fachbereich Chemie der Universität Marburg
D-3550 Marburg

List of Editors

ISBN 3-540-15784-0 Springer-Verlag Berlin Heidelberg New York Tokyo
ISBN 0-387-15784-0 Springer-Verlag New York Heidelberg Berlin Tokyo

The use of registered names, trademarks, etc. in this publication does not imply, even in the absence of a specific statement, that such names are exempt from the relevant protective laws and regulations and therefore free for general use.
Printing: Buch- und Offsetdruckerei H. Heenemann, Berlin. Bookbinding: Lüderitz & Bauer, Berlin.

2152/3020-543210

Dedicated to my coworkers

Preface

Titanium has been used to perform many kinds of reactions in organic and inorganic chemistry. The present book is concerned primarily with a new development in titanium chemistry which is useful in organic synthesis. In 1979/80 it was discovered that the titanation of classical carbanions using ClTiX$_3$ leads to species with reduced basicity and reactivity. This increases chemo-, regio- and stereoselectivity in reactions with organic compounds such as aldehydes, ketones and alkyl halides. Many new examples have been reported in recent times. Since the nature of the ligand X at titanium can be widely varied, the electronic and steric nature of the reagents is easily controlled. This helps in predicting the stereochemical outcome of many of the C—C bond forming reactions, but the trial and error method is still necessary in other cases. One of the ultimate objectives of chemistry is to understand correlations between structure and reactivity. Although this goal has not been reached in the area of organotitanium chemistry, appreciable progress has been made. A great deal of physical and computational data of organotitanium compounds described in the current and older literature (e.g., Ziegler-Natta type catalysts) has been reported by polymer, inorganic and theoretical chemists. It is summarized in Chapter 2 of this book, because some aspects are useful in understanding reactivity and selectivity of organo-titanium compounds in organic synthesis as described in the chapters which follow. Included are also novel reaction types, in addition to new applications of older reactions, e.g., TiCl$_4$ mediated additions of enol and allyl silanes to carbonyl compounds. Finally, limitations of the reactions are discussed throughout, and comparisons with other metals and methodologies are made. The literature up to early 1985 has been considered and some unpublished data.

The book was written purposely in the style of a progress report, because organotitanium chemistry has not reached the state of maturity as, for example, silicon chemistry has. Hopefully, it will contribute in reaching this goal.

Marburg, October 1985   M. T. Reetz

Table of Contents

Table of Contents

1. Introduction

1.1 Adjustment of Carbanion-Selectivity via Titanation

Carbanion chemistry represents an integral part of organic synthesis [1]. Not only simple Grignard- and alkyllithium compounds are useful in C—C bond formation, but also a host of heteroatom-substituted and resonance-stabilized species generated by such classical methods as halogen-metal exchange or deprotonation of CH-acidic organic compounds. Countless examples of addition reactions of carbonyl compounds (Grignard- and aldol-type), Michael additions, substitution processes involving alkyl halides or sulfonates as well as Wittig-like olefinations have been reported.

Inspite of the usefulness of these reactions, a number of problems persist. In most cases the reagents are extremely basic and reactive, which means that only a limited number of additional functional groups are tolerated [1]. Organic chemists have accepted this lack of chemoselectivity. Thus, no one is likely to devise a synthetic sequence in which such reactive species as lithium ester enolates or Grignard compounds are added to ketoaldehydes or other carbonyl compounds containing additional sensitive functionality. These reagents do not discriminate effectively between the different acceptor sites of a polyfunctional molecule, and product mixtures result.

The question of regioselectivity arises when the reagent or the substrate is ambident [1]. The problem of 1, 2 versus 1, 4 addition to α,β-unsaturated carbonyl compounds has been solved using alkyllithium compounds and cuprates [2], respectively, but general methods for regioselective control in ambident carbanions have not been devised [3]. In recent years enormous progress has been made in stereoselective C—C bond formation using carbanion chemistry, e.g., Meyers' oxazoline chemistry [4], aldol additions [5] and related reactions [6]. Nevertheless, in numerous other cases the success of the standard arsenal of reagents in stereoselective processes is mediocre [7]. Obviously, synthetic organic chemists need a wide variety of complementary methods.

This book describes a new principle in carbanion chemistry which is proving to be useful. In 1979/80 it was discovered that certain organo-titanium(IV) reagents behave much more selectively than their lithium or magnesium counterparts [8–14]. These observations led to the general working hypothesis that titanation of classical carbanions increases chemo-, regio- and stereoselectivity in reactions with carbonyl compounds, alkyl halides and other electrophiles [15–17]. The type of bond formation is not based on typical transition metal behavior such as oxidative coupling, β-hydride elimination or CO insertion. Rather, reaction types traditional to carbanion chemistry are most often involved.

1

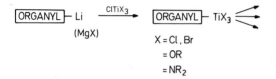

X = Cl , Br
= OR
= NR$_2$

By choosing the proper ligand X, two important parameters can be controlled in a predictable way [17]:
1) The electronic property at titanium, e.g., Lewis acidity, and
2) the steric environment around the metal.
For example, organyltitanium trichlorides RTiCl$_3$ are strong Lewis acids, a property which is important in chelation-controlled additions to chiral alkoxy carbonyl compounds (Chapter 5) or in alkylation reactions of S_N1-active alkyl halides (Chapter 7). Lewis acidity decreases drastically by going to organyltitanium tri-alkoxides RTi(OR')$_3$ or triamides RTi(NR'$_2$)$_3$, a prerequisite for non-chelation controlled additions to chiral alkoxy carbonyl compounds (Chapter 5). In all cases basicity of the reagents is considerably lower than that of RMgX or RLi [13–17]. Within the series RTi(OR')$_3$ or RTi(NR'$_2$)$_3$, the size of the R' substituent at oxygen or nitrogen determines the steric property of the reagent.
Further manipulation is possible using mixed ligand systems having two or three different X groups at titanium. For example, Lewis acidity can be modulated in a predictable way by stepwise replacement of chlorine by alkoxy ligands, the acidity decreasing in the following series:

$$RTiCl_3 > RTiCl_2(OR') > RTiCl(OR')_2 > RTi(OR')_3$$

Relevant to the use of mixed systems are ligands not mentioned thus far, for example, h^5-cyclopentadienyl groups. These exert such pronounced electron releasing effects, that Lewis acidity as well as reactivity with respect to addition reactions to carbonyl compounds decrease considerably. On the other hand, substituting alkoxy by methyl groups has the opposite effect (Chapters 2-4). For example, the rate of methyl addition to carbonyl groups increases dramatically in the following series [17, 18]:

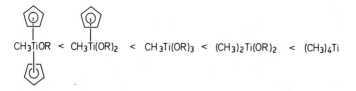

Not only the nature of the ligand, but also the number of ligands at Ti(IV) influences reactivity and selectivity. This can be achieved by generating "ate" complexes (Chapter 3), which display different selectivity than the neutral analogs RTi(OCHMe$_2$)$_3$:

RLi + Ti(OCHMe$_2$)$_4$ → RTi(OCHMe$_2$)$_4$Li
(MgX)   (MgX)

Clearly, there are many ways to produce tailor-made titanium reagents. In principle, controlled adjustment of carbanion-selectivity via titanation also allows for chiral modification (Chapter 5), either in the ligand system (e.g., *1*, *2*), at titanium (e.g., *3*) or as a combination of both (e.g., *4*).

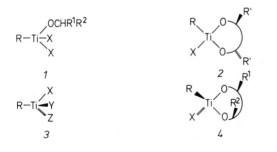

Although there are certain synthetic limitations (which will be described at relevant points in this book), experience in our laboratory and in others during the last five years has shown that organotitanium(IV) chemistry as outlined here is surprisingly versatile. Undoubtedly, further applications of the above principle will follow. Generally, the titanating agents $ClTiX_3$ are cheap. For example, $ClTi(OCHMe_2)_3$ is prepared quantitatively by mixing $TiCl_4$ and $Ti(OCHMe_2)_4$ in a ratio of $1:3$ [19]. $ClTi(NEt_2)_3$ is also made from $TiCl_4$ [20]. Preparation of these titanating agents on a large laboratory scale (e.g., 1 molar) poses no problems. Furthermore, workup of reactions involving addition or alkylation reactions of $RTiX_3$ does not lead to toxic materials.

$$1\ TiCl_4 + 3\ Ti(OCHMe_2)_4 \rightarrow 4\ ClTi(OCHMe_2)_3$$

$$1\ TiCl_4 + 3\ Et_2NLi \rightarrow 4\ ClTi(NEt_2)_3$$

A further aspect of organyltitanium compounds concerns the possibility of unusual reaction types, e.g., geminal dimethylation of ketones using $(CH_3)_2TiCl_2$ [21] (Chapter 7) or Wittig-type olefination employing $CH_2Br_2/Zn/TiCl_4$ [22] (Chapter 8).

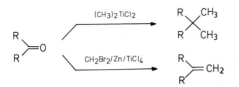

1.2 Other Uses of Titanium in Organic Chemistry

A number of useful processes involving titanium will not be treated in detail in this book [23]. Of paramount importance in this respect is the Ziegler-Natta polymerization of ethylene, propylene, and other α-olefins [24]. In the

1. Introduction

original process the heterogeneous catalyst is formed by mixing triethyl-aluminum and titanium tetrachloride. Although the precise mechanism of catalytic action is still a matter of controversy, one general and widely accepted aspect concerns coordination of the alkene to the active metal center prior to insertion into a titanium alkyl bond. This transition metal-like behavior is also operational in other, more active titanium-based catalysts which have been developed during the last two decades. Indeed, efficient polymerization catalysts need not contain titanium at all, other transition metals such as chromium and vanadium also being effective [24].

$$CH_2 = CH_2 \xrightarrow{\text{Et}_3\text{Al}/\text{TiCl}_4} \text{~~}\text{~}_n\text{~}$$

The enormous commercial importance of Ziegler-Natta polymerization spurred interest in synthesizing and characterizing various monomeric alkyl-titanium compounds [23], including CH_3TiCl_3 [25]. They were tested for catalytic activity in polymerization, but not for C—C bond forming processes in organic synthesis. Interestingly, some were shown to give a positive, others a negative Gilman test, e.g., $Ti(CH_3)_4$ [26] and CH_3TiCl_3 [23], respectively.

A large number of h^5-cyclopentadienyltitanium compounds have been prepared [23, 27], some for the purpose of testing their activity in various catalytic and stoichiometric processes such as CO insertion [28], olefin-metathesis [29], reduction [30], hydrometallation [31] and carbometallation [32]. Much of this important work has been reviewed [23, 27]. Sometimes $TiCl_4$ can be used in place of Cp-titanium compounds. Typical examples are shown below:

Cp₂Ti(Ph)₂ $\xrightarrow{\text{CO}}$ Cp₂Ti(CO)₂ + Ph–CO–Ph   [28e]

R—Br $\xrightarrow[\text{Me}_2\text{CHMgCl}]{\text{Cp}_2\text{TiCl}_2}$ R—H   (30c]

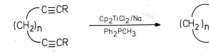   [30g]

   [30b]

         [30b]

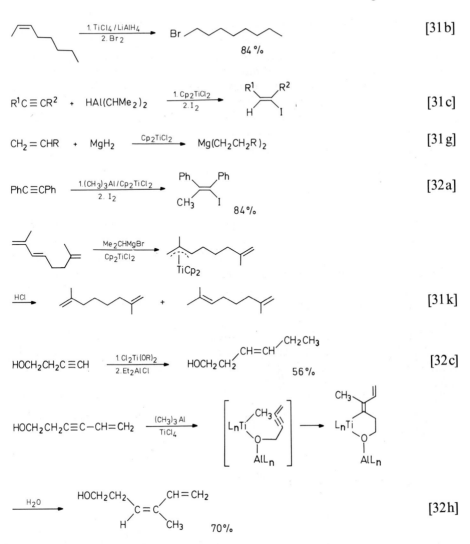

[31b]

[31c]

[31g]

[32a]

[31k]

[32c]

[32h]

An elegant application of low-valent titanium pertains to the reductive dimerization of ketones and aldehydes using the McMurry system (TiCl$_3$/LiAlH$_4$) or similar reagents [33]. The reaction can be stopped at the diol stage by proper choice of reagents and conditions. Corey has applied this to intramolecular C—C bond formation [34a].

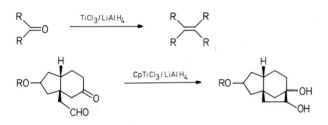

1. Introduction

McMurry has also cyclized keto-esters using TiCl$_3$/LiAlH$_4$, a process which delivers cyclic ketones in good yield [34b]. A synthetically interesting variation of pinacol type coupling concerns the use of aqueous TiCl$_3$, as demonstrated by Clerici and coworkers [33e–g]:

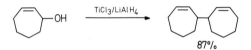

Aqueous TiCl$_3$ in basic medium is also a useful agent for other processes such as reduction of α-diketones, acyloins and cyano- and chloro-pyridines as well as the synthesis of allylic pinacols [33e–g].

Prior to the discovery of the above reactions, van Tamelan had shown that low-valent titanium affects the reductive coupling of benzylic and allylic alcohols [35]. Later an improved procedure was published [36].

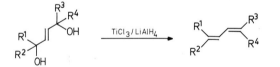

87%

Related are reductive eliminations of 1,2-glycols to olefins [37] as well as cyclization of 1,3-diols to cyclopropanes [38].

Along these lines, the titanium promoted reductive elimination of 2-ene-1,4-diols to 1,3-dienes according to Walborsky is of particular synthetic interest [39]. Since the starting diols are easily accessible (condensation of an aldehyde and/or ketone with acetylene followed by reduction), this 1,4-reductive elimination constitutes an attractive synthesis of dienes.

Titanium tetrachloride has found application in various areas of organic synthesis. An early use involves Friedel-Crafts alkylations, in which TiCl$_4$ is frequently more efficient than other Lewis acids such as AlCl$_3$ [40]. Trost has reported intramolecular variations involving sulfones [40b].

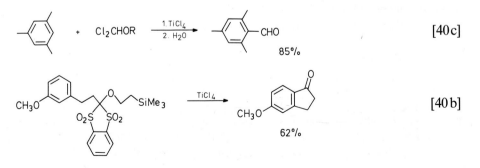

6

For some time Mukaiyama has utilized $TiCl_4$ in crossed-aldol reactions and Michael additions involving enol silanes [5d, 41].

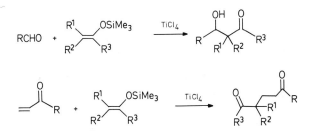

Allylsilanes undergo similar reactions, as shown by Sakurai and Hosomi [42]. Crotylsilanes react stereoselectively with $RCHO/TiCl_4$ [43]. In all of these reactions $TiCl_4$ activates the carbonyl component by Lewis acid/ Lewis base complexation. An interesting case of chirality transfer has been reported by Kumada [43b]. Fascinating examples of intramolecular allyl-silane and stannane additions have been described by Denmark and others, who studied the mechanism and stereoselectivity of such processes as a function of the Lewis acid (including $TiCl_4$) [43c, d].

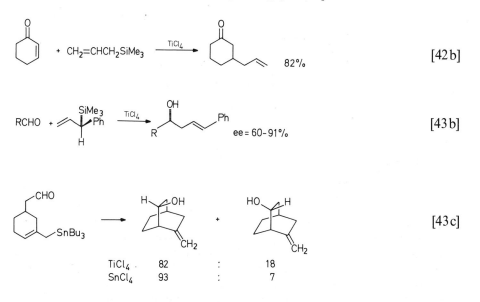

The α-alkylation of ketones via the corresponding enol silanes using S_N1-active alkyl halides is another recent development [44]. Such processes are of particular synthetic value because they are complementary to classical alkylations via reactions of lithium enolates with S_N2-active alkyl halides. Thus, none of the following reactions proceed classically with lithium eno-lates due to competing HX-eliminations. It should be pointed out that in special cases milder Lewis acids such as $SnCl_4$ or ZnX_2 are better suited [44].

1. Introduction

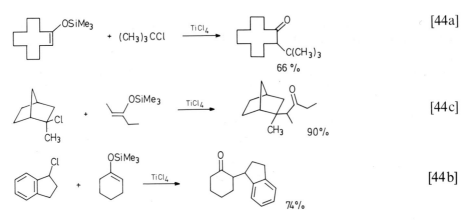

[44a]

66 %

[44c]

90%

[44b]

74%

TiCl$_4$ is often used to promote Diels-Alder reactions at low temperatures, particularly in efforts directed toward asymmetric induction. This important synthetic development goes back to the pioneering work of Walborsky concerning the TiCl$_4$ mediated addition of bis(menthyl)fumarate *5* to butadiene [45]. A Prelog-type of conformation *6* was postulated to account for the high stereoselectivity (de = 78%).

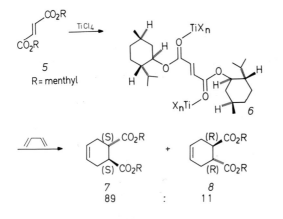

5
R= menthyl

6

7
89

8
11

:

Another milestone in the area of asymmetric Diels-Alder reactions was set by Corey, who showed that the acrylate *9* undergoes a highly stereoselective Diels-Alder reaction with the cyclopentadiene derivative *10* [46]. Later, Oppolzer generalized this method by employing *9* and its enantiomer in combination with other dienes and various Lewis acids [47]. The results (de up to 93%) were rationalized by assuming steric shielding on the part of the phenyl group.

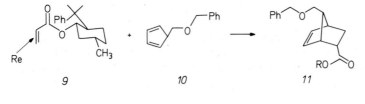

9

10

11

Even more efficient chiral auxiliaries have been reported by Oppolzer [48], Helmchen [49], Trost [50], Masamune [51] and others. Since a comprehensive review has appeared recently [48], only a few aspects are mentioned here. $TiCl_4$ and the milder $TiCl_2(OCHMe_2)_2$ are often the Lewis acids of choice, but in other cases $SnCl_4$, BF_3-etherate, $B(OAc)_3$ or R_2AlCl are better suited [48]. An interesting case in which $TiCl_4$ is particularly effective was reported by Helmchen [49], who treated the lactate 12 with $TiCl_4$ prior to Diels-Alder reaction with cyclopentadiene. The X-ray structure of the primary Lewis acid/Lewis base adduct 13 (see Chapter 2, Section 2.3) shows that it is one of the chlorine atoms around the octahedral titanium which shields the Re side of the olefinic double bond.

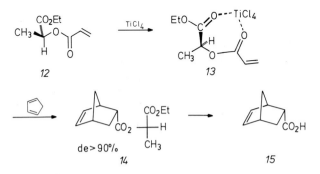

Diels-Alder reactions of heterodienophiles are sometimes catalyzed by Lewis acids [52]. Recently, a series of important papers by Danishefsky concerning the cycloaddition of siloxy dienes with aldehydes has appeared [53–56], e.g.:

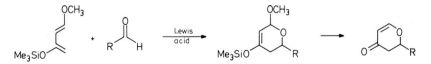

Useful Lewis acids include $ZnCl_2$, BF_3-etherate, $MgBr_2$, $Yb(fod)_3$ and $TiCl_4$. Besides the problem of regiocontrol, stereochemical aspects are involved, e.g., endo/exo-selectivity and diastereofacial selectivity in case of chiral aldehydes. Several definitive papers describing the synthetic solution to these problems as well as mechanistic studies and applications in natural products chemistry summarize the enormous progress made in this area [55–56]. It turns out that two extreme mechanisms must be considered:

1) a pericyclic addition similar to the classical Diels-Alder cycloaddition, and

2) a two step sequence initiated by a Mukaiyama type of aldol addition. Further mechanistic details have been discussed in terms of the nature of the diene, aldehyde, Lewis acid, and solvent [55a]. Sometimes the initial adduct has to be cyclized in an additional synthetic step by treatment with trifluoroacetic acid (TFA).

1. Introduction

One of the intriguing applications involves addition reactions of chiral α- and β-alkoxy aldehydes [55b]. In this case the nature of the Lewis acid determines the stereochemical outcome in a predictable way. For example, the diene *16* reacts with the β-alkoxy aldehyde *17* in the presence of MgBr$_2$ via a pericyclic pathway to produce the so-called trans-chelation-controlled product. In contrast, BF$_3$ initiates a Mukaiyama-like addition, resulting in adducts which have to be cyclized with TFA. The sequence is of little value due to the low degree of steroselectivity. TiCl$_4$ is the Lewis acid of choice because a single product *18* is formed [55b]. This is in line with previous results concerning asymmetric induction in the TiCl$_4$ mediated aldol addition of enol and allylsilanes with chiral β-alkoxy aldehydes [57]. TiCl$_4$ forms a six-membered chelate which is attacked stereoselectively from the least hindered side [58] (see Chapter 5, Section 5.2.2 for a detailed discussion of such reactions).

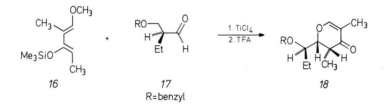

| 16 | 17 | 18 |

R=benzyl

Other applications of TiCl$_4$ as a Lewis acid in combination with silicon compounds are discussed in recent reviews [59]. Finally, TiCl$_4$-mediated condensation reactions such as Knoevenagel reactions of active methylene compounds with aldehydes are useful processes [60]. Also, it has been reported that enamine formation proceeds particularly well in the presence of TiCl$_4$ [61]. Some of these reactions will be considered in the chapters which follow.

Titanium tetraalkoxides of the general formula Ti(OR)$_4$ have been used in industry for a long time in various ways, including esterification and transesterification [62]. Recently, the latter has been applied by Seebach [63] and Steglich [64] to sensitive substrates which decompose or racemize under the conditions of classical methods. These mild reactions are likely to gain importance in peptide chemistry [64a]. It is interesting to note that carboxylic acid esters are transesterified chemoselectively in the presence of phosphonic acid esters [65].

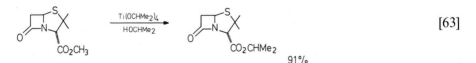

[63]

91%

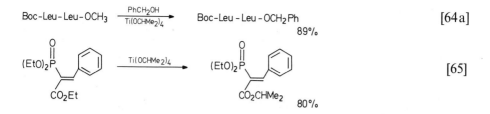

$$\text{Boc-Leu-Leu-OCH}_3 \xrightarrow[\text{Ti(OCHMe}_2)_4]{\text{PhCH}_2\text{OH}} \text{Boc-Leu-Leu-OCH}_2\text{Ph} \quad 89\% \qquad [64a]$$

[65]

A very exciting application of Ti(OCHMe$_2$)$_4$ concerns the Sharpless enantioselective epoxidation of allyl alcohols using *tert*-butylhydroperoxide and tartaric acid esters [66]. Of prime importance is the fact that the process is often substrate independent. Thus, the chirality of the catalyst overrides the effect of any chiral center which may be present in the allyl alcohol! Several reviews covering synthetic and mechanistic aspects have appeared [66]. The titanium-mediated stereoselective ring-opening of certain epoxides is also of synthetic interest [67a], including those in which the azide function is introduced under mild conditions [67b].

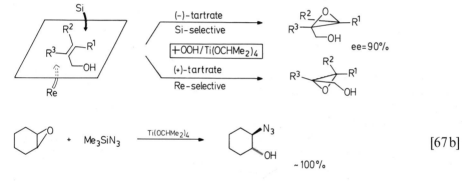

[67b]

Finally, the photolysis of titanium compounds (Chapter 2) is noteworthy. The TiCl$_4$ mediated photolysis of ketones in the presence of methanol according to Sato is of synthetic interest [68]. The reaction is equivalent to the addition of a CH$_2$OH fragment to the carbonyl function. This interesting electron transfer process has been applied in the synthesis of the pheromone frontalin [68].

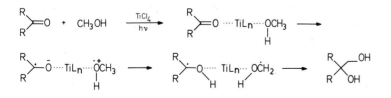

This brief survey is not meant to be exhaustive. Further information, e.g., concerning the role of titanium in nitrogen fixation [69] and in other processes can be found in monographs and reviews [23, 27, 70].

References

1. a) Stowell, J. C.: Carbanions in Organic Synthesis, Wiley, N.Y. 1979; b) Bates, R. B., Ogle, C. A.: Carbanion Chemistry, Springer-Verlag, Berlin 1983; c) Caine, D.: in Carbon—Carbon Bond Formation, (Augustine, R. L., editor), Vol. I, Marcel Dekker 1979; d) Wakefield, B. J.: "The Chemistry of Organolithium Compounds", Pergamon Press N.Y. 1974.
2. a) Posner, G. H.: Org. React. *19*, 1 (1972); Org. React. *22*, 253 (1975); b) Lipshutz, B. H., Wilhelm, R. S., Kozlowski, J. A.: Tetrahedron *40*, 5005 (1984).
3. a) Gompper, R., Wagner, H. U.: Angew. Chem. *88*, 389 (1976); Angew. Chem., Int. Ed. Engl. *15*, 321 (1976); b) review of α, γ-reactivity of substituted allylic organometallic reagents: Biellmann, J.-F., Ducep, J. B.: Org. React. *27*, 1 (1982).
4. a) Meyers, A. I.: Acc. Chem. Res. *11*, 375 (1978); b) Barner, B. A., Meyers, A. I.: J. Am. Chem. Soc. *106*, 1865 (1984).
5. a) Evans, D. A., Nelson, J. V., Taber, T. R.: Top. Stereochem. *13*, 1 (1982); b) Heathcock, C. H.: in "Asymmetric Synthesis", Morrison, J. D. (editor), Vol. III, Academic Press, N.Y. 1984); c) Masamune, S., Choy, W.: Aldrichim. Acta *15*, 47 (1982); d) Mukaiyama, T.: Org. React. *28*, 203 (1982).
6. Reviews: a) Hoffmann, R. W.: Angew. Chem. *94*, 569 (1982); Angew. Chem., Int. Ed. Engl. *21*, 555 (1982); b) Yamamoto, Y., Maruyama, K.: Heterocycles *18*, 357 (1982).
7. a) Morrison, J. D., Mosher, H. S.: "Asymmetric Organic Reactions", Prentice Hall, Englewood Cliffs, N.J. 1971; b) Bartlett, P. A.: Tetrahedron *36*, 3 (1980); c) Martens, J.: Top. Curr. Chem. *125*, 165 (1984).
8. Reetz, M. T., Westermann, J., Steinbach, R.: Experiments performed at the University of Bonn, 1979. The early part of our work was first communicated in lectures presented at the Universität München (November 10, 1979). Bayer AG, Krefeld (November 15, 1979) and Max-Planck-Institut für Kohleforschung, Mülheim (December 10, 1979).
9. Reetz, M. T., Westermann, J., Steinbach, R.: Angew. Chem. *92*, 931 (1980); Angew. Chem., Int. Ed. Engl. *19*, 900 (1980).
10. Reetz, M. T., Westermann, J., Steinbach, R.: Angew. Chem. *92*, 933 (1980); Angew. Chem., Int. Ed. Engl. *19*, 901 (1980).
11. Reetz, M. T., Wenderoth, B., Peter, R., Steinbach, R., Westermann, J.: J. Chem. Soc., Chem. Commun. *1980*, 1202.
12. a) Reetz, M. T., Steinbach, R., Westermann, J., Peter, R.: Angew. Chem. *92*, 1044 (1980); Angew. Chem., Int. Ed. Engl. *19*, 1011 (1980).
13. a) Westermann, J.: Diplomarbeit, Univ. Bonn 1980; b) Steinbach, R.: Diplomarbeit, Univ. Bonn 1980; c) Peter, R.: Diplomarbeit, Univ. Marburg 1980; d) Wenderoth, B.: Diplomarbeit, Univ. Marburg 1980.
14. See also a) Weidmann, B., Seebach, D.: Helv. Chim. Acta *63*, 2451 (1980); b) Weidmann, B., Seebach, D.: Angew. Chem. *95*, 12 (1983); Angew. Chem., Int. Ed. Engl. *22*, 31 (1983).
15. Reetz, M. T., Steinbach, R., Wenderoth, B., Westermann, J.: Chem. Ind. *1981*, 541.
16. Reetz, M. T.: Nachr. Chem. Techn. Lab. *29*, 165 (1981).
17. a) Reetz, M. T.: Top. Curr. Chem. *106*, 1 (1982); b) Reetz, M. T.: Pure Appl. Chem. *57*, 1781 (1985).
18. Reetz, M. T. et al.: unpublished results 1982–1985.

19. Reetz, M. T., Steinbach, R., Kesseler, K.: Angew. Chem. *94*, 872 (1982); Angew. Chem., Int. Ed. Engl. *21*, 864 (1982); Angew. Chem. Supplement *1982*, 1899.

20. Reetz, M. T., Urz, R., Schuster, T.: Synthesis *1983*, 540.

21. Reetz, M. T., Westermann, J., Kyung, S. H.: Chem. Ber. *118*, 1050 (1985).

22. a) Takai, K., Hotta, Y., Oshima, K., Nozaki, H.: Tetrahedron Lett. *19*, 2417 (1978); b) Takai, K., Hotta, Y., Oshima, K., Nozaki, H.: Bull. Chem. Soc. Jap. *53*, 1698 (1980); c) Lombardo, L.: Tetrahedron Lett. *23*, 4293 (1982).

23. Reviews of Ti(II), Ti(III) and Ti(IV) Compounds: a) Gmelin Handbuch, "Titan-Organische Verbindungen", Part 1 (1977), Part 2 (1980), Part 3 (1984), Part 4 (1984), Springer-Verlag, Berlin; b) Segnitz, A.: in Houben-Weyl-Müller, "Methoden der Organischen Chemie", Vol. 13/7, p. 261, Thieme-Verlag, Stuttgart 1975); c) Wailes, P. C., Coutts, R. S. P., Weigold, H.: "Organometallic Chemistry of Titanium, Zirconium and Hafnium", Academic Press, N.Y. 1974; d) Labinger, J. A.: J. Organomet. Chem. *227*, 341 (1982); e) Bottrill, M., Gavens, P. D., Kelland, J. W., McMeeking, J.: in "Comprehensive Organometallic Chemistry", (Wilkinson, G., Stone, F. G. A., Abel, E. W., editors), Chapter 22, Pergamon Press, Oxford (1982).

24. a) Sinn, H., Kaminsky, W.: Adv. Organomet. Chem. *18*, 99 (1980); b) Pino, P., Mülhaupt, R.: Angew. Chem. *92*, 869 (1980); Angew. Chem., Int. Ed. Engl. *19*, 857 (1980); c) see also lit. [23e].

25. a) Beermann, C., Bestian, H.: Angew. Chem. *71*, 618 (1959); b) for other syntheses as well physical and chemical properties, see ref. [23] as well as Chapter 2.

26. a) Clauss, K., Beermann, C.: Angew. Chem. *71*, 627 (1959); b) Berthold, H. J., Groh, G.: Z. Anorg. Allg. Chem. *319*, 230 (1963); c) for further details concerning this compound, see ref. [23] and Chapter 2.

27. Review: Pez, G. P., Armor, J. N.: Adv. Organomet. Chem. *19*, 1 (1981).

28. See for example: a) Murray, J. G.: J. Am. Chem. Soc. *83*, 1287 (1961); b) Fachinetti, G., Floriani, C.: Organomet. Chem. *71*, C5 (1974); c) McDermott, J. X., Whitesides, G. M.: J. Am. Chem. Soc. *96*, 947 (1974); d) Siegert, F. W., De Liefde Meijer, H. J.: J. Organomet. Chem. *20*, 141 (1969); e) Masai, H., Sonogashira, K., Hagihara, N.: Bull. Chem. Soc. Jap. *41*, 750 (1968).

29. a) Lee, J. B., Gajda, G. J., Schaefer, W. P., Howard, T. R., Ikariya, T., Straus, D. A., Grubbs, R. H.: J. Am. Chem. Soc. *103*, 7358 (1981); b) Grubbs, R. H.: Prog. Inorg. Chem. *24*, 1 (1979).

30. See for example: a) Brown, H. C., Subba Rao, B. C.: J. Am. Chem. Soc. *78*, 2582 (1956); b) Kano, S., Tanaka, Y., Sugino, E., Hibino, S.: Synthesis *1980*, 695, 741; c) Rilatt, J. A., Kitching, W.: Organometallics, *1*, 1089 (1982); d) Lipshutz, B. H., Morey, M. C.: Tetrahedron Lett. *25*, 1319 (1984); e) Welch, S. C., Walters, M. E.: J. Org. Chem. *43*, 4797 (1978); f) Nelson, T. R., Tufariello, J. J.: J. Org. Chem. *40*, 3159 (1975); g) Nugent, W. A., Calabrese, J. C.: J. Am. Chem. Soc. *106*, 6422 (1984).

31. See for example: a) Isagawa, K., Sano, H., Hattori, M., Otsuji, Y.: Chem. Lett. *1979*, 1069; b) Sato, F., Ishikawa, H., Sato, M.: Tetrahedron Lett. *22*, 85 (1981); c) Ashby, E. C., Noding, S. R.: J. Organomet. Chem. *177*, 117 (1979); d) Eisch, J. J., Galle, J. E.: J. Organomet. Chem. *160*, C8 (1978); e) Fell, B., Asinger, F., Sulzbach, R. A.: Chem. Ber. *103*, 3830 (1970); f) Lee, H. S., Isagawa, K., Otsuji, Y.: Chem. Lett. *1984*, 363; g) Bogdanovic, B., Schwickardi, M., Sikorsky, P.: Angew. Chem. *94*, 206 (1982); Angew. Chem., Int. Ed. Engl. *21*, 199 (1982); Angew. Chem. Supplement *1982*, 457; h) Lehmkuhl, H., Fustero, S.: Liebigs Ann. Chem. *1980*, 1353.

32. See for example: a) Negishi, E.: Pure Appl. Chem. *53*, 2333 (1981); b) Brown, D. C., Nichols, S. A., Gilpin, A. B., Thompson, D. W.: J. Org. Chem. *44*, 3457 (1979); c) Tweedy, H. E., Coleman, R. A., Thompson, D. W.: J. Organomet. Chem. *129*, 69 (1977); d) Ashby, E. C., Noding, S. A.: J. Org. Chem. *45*, 1035 (1980); e) Carr, D. B., Schwartz, J.: J. Am. Chem. Soc. *101*, 3521 (1979); f) Schiavelli, M. D., Plunkett, J. J., Thompson, D. W.: J. Org. Chem. *46*, 807 (1981); g) Wilke, G.: J. Organomet. Chem. *200*, 349 (1980); h) Zitzelberger, T. J., Schiavelli, M. D., Thompson, D. W.: J. Org. Chem. *48*, 4781 (1983); for comparisons with zirconium mediated reactions, see [32a] and i) Yoshida, T.: Chem. Lett. *1982*, 293, as well as j) Schwartz, J.: J. Organomet. Chem., Library *1*, 461 (1976); k) reactions of h^3-allyltitanium(III) compounds with ethylene: Lehmkuhl, H., Janssen, E., Schwickardi, R.: J. Organomet. Chem. *258*, 171 (1983).
33. a) McMurry, J. E.: Acc. Chem. Res. *7*, 281 (1974); b) McMurry, J. E.: Acc. Chem. Res. *16*, 405 (1983); c) Dams, R., Malinowski, M., Westdorp, I., Geise, H. Y.: J. Org. Chem. *47*, 248 (1982); d) Ledon, H., Tkachenko, I., Young, D.: Tetrahedron Lett. *20*, 173 (1979); e) Clerici, A., Porta, O.: J. Org. Chem. *47*, 2852 (1982); f) Clerici, A., Porta, O.: J. Org. Chem. *48*, 1690 (1983); g) Clerici, A., Porta, O.: J. Org. Chem. *50*, 76 (1985).
34. a) Corey, E. J., Danheiser, R. L., Chandrasekaran, S.: J. Org. Chem. *41*, 260 (1976); b) McMurry, J. E., Miller, D. D.: J. Am. Chem. Soc. *105*, 1660 (1983).
35. a) Van Tamelen, E. E., Schwartz, M. A.: J. Am. Chem. Soc. *87*, 3277 (1965); b) Sharpless, K. B., Hanzlik, R. P., van Tamelen, E. E.: J. Am. Chem. Soc. *90*, 209 (1968).
36. McMurry, J. E., Silvestri, M.: J. Org. Chem. *40*, 2687 (1975).
37. a) McMurry, J. E., Fleming, M. P., Kees, K. L., Krepski, L. R.: J. Org. Chem. *43*, 3255 (1978); b) review of deoxygenation of vicinal diols: Block, E.: Org. React. *30*, 457 (1984).
38. a) Baumstark, A. L., McCloskey, C. J., Tolson, T. J., Syriopoulos, G. T.: Tetrahedron Lett. *18*, 3003 (1977); b) Walborsky, H. M., Pass Murari, M.: J. Am. Chem. Soc. *102*, 426 (1980).
39. Walborsky, H. M., Wüst, H. H.: J. Am. Chem. Soc. *104*, 5807 (1982).
40. See for example: a) Cullinane, N. M., Leyshon, D. M.: J. Chem. Soc. *1954*, 2942; b) intramolecular Friedel-Crafts reaction of sulfones: Trost, B. M., Quayle, P.: J. Am. Chem. Soc. *106*, 2469 (1984); c) Rieche, A., Groß, H., Höft, E.: Chem. Ber. *93*, 88 (1960).
41. Mukaiyama, T.: Angew. Chem. *89*, 858 (1977); Angew. Chem., Int. Ed. Engl. *16*, 817 (1977).
42. a) Hosomi, A., Sakurai, H.: Tetrahedron Lett. *1976*, 1295; b) Hosomi, A., Sakurai, H.: J. Am. Chem. Soc. *99*, 1673 (1977); c) review of allylsilane additions: Sakurai, H.: Pure Appl. Chem. *54*, 1 (1982); comparison with other Lewis acids: d) Calas, R., Dunogues, J., Deleris, G., Pisciotti, F.: J. Organomet. Chem. *69*, C15 (1974); e) Abel, E. W., Rowley, R. J.: J. Organomet. Chem. *84*, 199 (1975).
43. a) TiCl$_4$-mediated stereoselective addition of crotylsilanes to aldehydes: Hayashi, T., Kabeta, K., Hamachi, I., Kumada, M.: Tetrahedron Lett. *24*, 2865 (1983); b) Hayashi, T., Konishi, M., Kumada, M.: J. Org. Chem. *48*, 281 (1983); c) Denmark, S. E., Weber, E. J.: J. Am. Chem. Soc. *106*, 7970 (1984); d) Mikami, K., Maeda, T., Kishi, N., Nakai, T.: Tetrahedron Lett.: *25*, 5151 (1984), and literature cited therein concerning previous studies.

44. a) Reetz, M. T., Maier, W. F., Chatziiosifidis, I., Giannis, A., Heimbach, H., Löwe, U.: Chem. Ber. *113*, 3741 (1980); b) Reetz, M. T., Walz, P., Hübner, F., Hüttenhain, S. H., Heimbach, H., Schwellnus, K.: Chem. Ber. *117*, 322 (1984); c) Review of Lewis acid mediated α-alkylations of carbonyl compounds using S_N1-active alkylating agents: Reetz, M. T.: Angew. Chem. *94*, 97 (1982); Angew. Chem., Int. Ed. Engl. *21*, 96 (1982); see also d) Fleming, I.: Chimia *34*, 265 (1980); e) Brownbridge, P.: Synthesis *1983*, 1, 85.

45. a) Walborsky, H. M., Barash, L., Davis, T. C.: Tetrahedron *19*, 2333 (1963); b) see also Sauer, J., Kredel, J.: Tetrahedron Lett. *7*, 6359 (1966).

46. Corey, E. J., Ensley, H. E.: J. Am. Chem. Soc. *97*, 6908 (1975).

47. Oppolzer, W., Kurth, M., Reichlin, D., Moffatt, F.: Tetrahedron Lett. *22*, 2545 (1981).

48. Oppolzer, W.: Angew. Chem. *96*, 840 (1984); Angew. Chem., Int. Ed. Engl. *23*, 876 (1984).

49. a) Poll, T., Helmchen, G., Bauer, B.: Tetrahedron Lett. *25*, 2191 (1984); b) Poll, T., Metter, J. O., Helmchen, G.: Angew. Chem. *97*, *116* (1985): Angew. Chem., Int. Ed. Engl. *24*, 112 (1985).

50. Trost, B. M., O'Krongly, D., Belletire, J. C.: J. Am. Chem. Soc. *102*, 7595 (1980).

51. Choy, W., Reed, L. A., Masamune, S.: J. Org. Chem. *48*, 1139 (1983).

52. Review of the hetero-Diels-Alder reaction: Weinreb, S. M., Staib, R. R.: Tetrahedron *38*, 3087 (1982).

53. Danishefsky, S., Kerwin, J. F., Kobayashi, S.: J. Am. Chem. Soc. *104*, 358 (1982).

54. Danishefsky, S. J., Pearson, W. H., Harvey, D. F.: J. Am. Chem. Soc. *106*, 2456 (1984).

55. a) Danishefsky, S. J., Larson, E., Askin, D., Kato, N.: J. Am. Chem. Soc. *107*, 1246 (1985); b) Danishefsky, S. J., Pearson, W. H., Harvey, D. F., Maring, C. J., Springer, J. P.: J. Am. Chem. Soc. *107*, 1256 (1985).

56. a) Danishefsky, S. J., Maring, C. J.: J. Am. Chem. Soc. *107*, 1269 (1985); b) Danishefsky, S. J., Larson, E., Springer, J. P.: J. Am. Chem. Soc. *107*, 1274 (1985); c) Danishefsky, S. J., Pearson, W. H., Segmüller, B. E.: J. Am. Chem. Soc. *107*, 1280 (1985); d) Danishefsky, S. J., Uang, B. J., Quallich, G.: J. Am. Chem. Soc. *107*, 1285 (1985).

57. Reetz, M. T., Jung, A.: J. Am. Chem. Soc. *105*, 4833 (1983).

58. Review of chelation- and non-chelation-controlled additions to chiral α- and β-alkoxy carbonyl compounds: Reetz, M. T.: Angew. Chem. *96*, 542 (1984); Angew. Chem., Int. Ed. Engl. *23*, 556 (1984).

59. a) Colvin, E.: "Silicon in Organic Synthesis", Butterworths, London 1981; b) Weber, W. P.: "Silicon Reagents for Organic Synthesis", Springer-Verlag, Berlin 1983.

60. a) Lehnert, W.: Tetrahedron *30*, 301 (1974); b) Lehnert, W.: Tetrahedron Lett. *11*, 4723 (1970); c) Mukaiyama, T.: Pure Appl. Chem. *54*, 2455 (1982).

61. White, W. A., Weingarten, H.: J. Org. Chem. *32*, 213 (1967).

62. "Titansäureester", Information Pamphlet, Dynamit Nobel AG, Troisdorf, Federal Republic of Germany.

63. a) Seebach, D., Hungerbühler, E., Naef, R., Schnurrenberger, P., Weidmann, B., Züger, M.: Synthesis *1982*, 138; b) Seebach, D. in: "Modern Synthetic Methods", (Scheffold, editor), Vol. III, Salle Verlag Franfurt; Verlag Sauerländer Aarau, p. 217, 1983.

64. a) Rehwinkel, H., Steglich, W.: Synthesis *1982*, 826; b) Burke, S. D., Fobare, W. F., Pacofsky, G. J.: J. Org. Chem. *48*, 5221 (1983).

65. Peter, R.: Dissertation, Univ. Marburg 1983.
66. a) Katsuki, T., Sharpless, K. B.: J. Am. Chem. Soc. *102*, 5974 (1980); reviews: b) Sharpless, K. B., Woodard, S. S., Finn, M. G.: Pure Appl. Chem. *55*, 1823 (1983); c) Sharpless, K. B.: Proceedings of the Robert A. Welch Foundation Conferences on Chemical Research XXVII: Stereospecificity in Chemistry and Biochemistry, 1984; d) crystal structures of two catalysts: Williams, I. D., Pedersen, S. F., Sharpless, K. B., Lippard, S. J.: J. Am. Chem. Soc. *106*, 64 (1984).
67. a) Morgans, D. J., Sharpless, K. B.: J. Am. Chem. Soc. *103*, 462 (1981); b) Blandy, C., Choukroun, R., Gervais, D.: Tetrahedron Lett. *24*, 4189 (1983).
68. Sato, T., Kaneko, H., Yamaguchi, S.: J. Org. Chem. *45*, 3778 (1980).
69. a) Chatt, J., Dilworth, J. R., Richards, R. L.: Chem. Rev. *78*, 589 (1978); b) Vol'pin, M. E.: J. Organomet. Chem. *200*, 319 (1980).
70. a) Davies, S. G.: "Organotransition Metal Chemistry: Applications to Organic Synthesis", Pergamon Press, Oxford, 1982; b) Negishi, E.: "Organometallics in Organic Synthesis", Vol. I, Wiley, N.Y. 1980); c) Kochi, J. K.: "Organometallic Mechanisms and Catalysis", Academic Press, N.Y. 1978; d) Alper, H.: "Transition Metal Organometallics in Organic Synthesis", Vol. I and II, Academic Press, N.Y. 1976 and 1978.

2. Synthesis and Properties of Some Simple Organotitanium Compounds

Before embarking on a detailed discussion of the titanation of carbanions, synthetic and physical organic aspects of several typical organotitanium(IV) compounds shall be surveyed, including thermal stability, aggregation state, X-ray structural data and bond energies. Some of this information is useful in understanding reactivity and selectivity in reactions with organic substrates. Low valent Ti(II) and Ti(III) shall be mentioned only on passing; the interested reader is referred to reviews [1].

2.1 Synthesis and Stability

2.1.1 Historical Aspects

The history of organotitanium chemistry is fascinating, beginning with a long series of futile synthetic attempts. From todays perspective it is clear that most of these failures have to do with two aspects which should be kept in mind when working with organotitanium reagents. Firstly, early attempts were directed toward preparing tetra-alkyl or tetra-aryl derivatives of the structure TiR_4, analogous to the previously prepared silanes SiR_4, zinc compounds ZnR_2 or lead derivatives PbR_4. However, whereas the latter are distillable, it is now known that the titanium analogs are generally thermally very unstable, in contrast to many monomethyl or aryl compounds, e.g., CH_3TiCl_3, $CH_3Ti(OCHMe_2)_3$ or $RTi(NEt_2)_3$ [1].

Thus, the early research groups were thinking too much in terms of analogy and were in fact taking approaches which were least likely to suceed. Secondly, the initial efforts generally involved temperatures which are too high either during the actual synthesis or during workup (e.g., distillation at $> 100\ °C$).

The first experiments concerning the synthesis of alkyltitanium reagents were performed by Cahours in 1861 [2]. He reacted $TiCl_4$ with $ZnEt_2$ and tried to isolate $TiEt_4$. However, only black tarry material was formed. Also, elemental titanium failed to react with methyl iodide [2].

$$ZnEt_2 + TiCl_4 \rightarrow TiEt_4$$

Later, the reaction with diethylzinc was repeated under different conditions by other groups. Schumann observed the formation of reduced titanium as well as unidentified gases [3]. Paterno, upon attempting to distill the presumed TiEt$_4$, made similar observations and also isolated an oil which he identified as *n*-octane [4]. According to Levy, elemental titanium does not react with a variety of alkyl iodides, while diethylmercury combines with TiCl$_4$ (at 100 °C!) to yield EtHgCl, TiCl$_3$ and unidentified gases [5].

During this century, it was not until 1924 that further experimentation was reported. Upon reacting an excess of phenylmagnesium bromide with TiCl$_4$, Challenger notes [6]: "After remaining some days at ordinary temperature, the ether (A) and a dark oil (B) were decanted, while a black, viscid mass (C) remained. No trace of a phenyl compound of titanium could be isolated from these products". Shortly thereafter other groups also reported negative results, including the reaction of phenylmagnesium bromide with titanium trichloride at 180 °C [7].

The first systematic studies pertaining to the synthesis and attempted characterization of organotitanium compounds were done by Gilman and Jones [8]. The reaction of diphenylmercury with titanium powder at 130 °C for 12 days resulted in 98 % recovery of the starting components. In contrast, lively reactions of TiCl$_4$ or Ti(OEt)$_4$ with *n*-butyl- and phenyl-lithium were observed. Unfortunately, three or four equivalents of lithium reagent per titanium compound were (again) employed, which in fact thwarted isolation and characterization. For example, the reaction of TiCl$_4$ with four equivalents of *n*-butyllithium at -10 °C was reported to result in a black resinous material [8]. A similar reaction with phenyllithium afforded large amounts of diphenyl and resulted in the reduction of Ti(IV) to Ti(III). From todays viewpoint the most interesting reaction concerns the addition of four equivalents of phenyllithium to Ti(OEt)$_4$. The orange crystals which precipitated were reported to burn spontaneously in air, to react violently with water and to give a positive Gilman test (!) [8]. Apparently, the compound is not pure tetraphenyltitanium, because Gilman and Jones note that it contains lithium and halogen. Thus, some sort of an ate complex is more likely, although this has not been cleared up to date.

$$\text{Ti(OEt)}_4 + 4\,\text{PhLi} \xrightarrow[\text{ether}]{0\,°C} \text{orange crystals}$$

2.1.2 Mono-Aryl- and Alkyltitanium Compounds

In 1953 Herman and Nelson reported the first unambiguous synthesis and characterization of an organotitanium compound having a Ti—C σ-bond [9]. They reacted tetraisopropoxytitanium (*1*) with phenyllithium (containing LiBr) and treated the structurally undefined adduct with TiCl$_4$, obtaining tri-isopropoxyphenyltitanium (*2*) in an overall yield of 40%.

$$\text{Ti(OCHMe}_2)_4 + \text{PhLi(LiBr)} \xrightarrow{\text{ether}} \text{PhTi(OCHMe}_2)_4\text{LiBrOEt}_2$$
$$\textit{1}$$

$$\xrightarrow{\text{TiCl}_4} \text{PhTi(OCHMe}_2)_3$$
$$\textit{2}$$

Later, considerably improved procedures using the more convenient titanating agent chlorotriisopropoxytitanium (*3*) were developed. For example, salt-free phenyllithium was added to *3* in ether under nitrogen at −10 °C, lithium chloride filtered off and the solvent evaporated to yield 92% of the yellow crystalline compound *2* having a melting point of 88–90 °C [10]. The titanating agent *3* was made by treating *1* with acetylchloride [10]. A more convenient synthesis involves quantitative reaction of $TiCl_4$ with *1* [11].

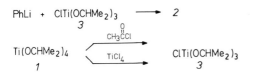

The phenyl compound *2* is stable in the dark at 10 °C for months, but decomposes rapidly if heated above its melting point (>90 °C) to form violet colored Ti(III) species and diphenyl [1, 10]. This pyrolytic property has been exploited in the polymerization of styrene and other olefines, but the compound appears not to be a particularly active polymerization catalyst [12]. Somewhat higher efficiency is achieved by exposition to light, *2* decomposing to $Ti(OCHMe_2)_3$ and phenyl radicals in a first order rate process [12]. The lesson to be learned for synthetic organic chemists is that organotitanium reagents should not be exposed to direct sunlight for long periods of time.

Compound *2* is sensitive to air (oxygen) and to moisture, phenol and benzene being formed, respectively [9, 13]. Thus, it (as well as most other organotitanium compounds) should be handled under an inert gas atmosphere.

Herman and Nelson also studied certain methyl, *n*-butyl and ethynyl derivatives by adding the corresponding lithium or magnesium reagents to tetrabutoxytitanium [14]. However, $TiCl_4$ was not added to the solution as in case of the phenyl compound. Thus, it is not clear whether the authors obtained adducts having Ti—C bonds (ate complexes) or whether compounds of the type $RTi(OBu)_3$ were formed. Other than a positive Gilman test, no characterization or isolation was attempted. Nevertheless, thermal stability of the adducts with respect to Ti(III) formation (i.e., decomposition) was reported to be as follows [14]:

phenyl > *p*-anisyl > ethynyl > methyl > *n*-butyl

Two other events in the early fifties spurred interest in organotitanium compounds: The discovery of Ziegler-Natta catalysts [15] and the characterization of ferrocene [16]. On the one hand, research efforts centered around the preparation of new organotitanium compounds for olefin polymerization; on the other hand, many inorganic chemists became interested in preparing h^5-cyclopentadienyl compounds containing titanium [17].

Several methyltitanium compounds CH_3TiX_3 were first prepared at Farbwerke Hoechst [18], further examples and improvements followed later. Triisopropoxymethyltitanium (4) is accessible by reacting methyllithium with chlorotriisopropoxytitanium (3), separation from the inorganic material and distillation at 50 °C/0.01 torr (~95% yield) [19, 20]. The yellow compound is best stored in the refrigerator in pure form (which may cause Of course, for synthetic purposes (Chapter 3 and 5) distillation is not partial crystallization) or in solution (e.g., ether, THF, pentane, toluene) [21]. necessary, i.e., an in situ reaction mode is equally, or even more, convenient [21, 22].

$$CH_3Li \overset{3}{\rightarrow} CH_3Ti(OCHMe_2)_3$$
$$4$$

The trichloro derivative 5 is accessible in high yield by treating $TiCl_4$ with a variety of methylmetal compounds such as $Al(CH_3)_3$ [18], $AlCl_2CH_3$ [18], $Cd(CH_3)_2$ [18], $Zn(CH_3)_2$ [18, 23, 24], $Pb(CH_3)_4$ [25], CH_3MgBr [26], or CH_3Li [27]. If an ether-free solution is needed, quantitative methylation of $TiCl_4$ using $Zn(CH_3)_2$ in pentane or hexane constitutes an excellent procedure [24]. Although pure $Zn(CH_3)_2$ is pyrophoric, CH_2Cl_2 solutions are much less dangerous and can be handled like n-butyllithium solutions [28]; thus, 5 can easily be prepared in this solvent [29]. The use of CH_3Li involves ethereal solutions which leads to the bis-etherate of CH_3TiCl_3 (6). Both, the free and the complexed form are synthetically useful, depending upon the type of C—C bond forming reaction desired (Chapters 3, 5, 7).

$$TiCl_4 \quad \overset{1/2 Zn(CH_3)_2}{\underset{CH_2Cl_2}{\longrightarrow}} \quad CH_3TiCl_3 \atop 5$$
$$\overset{CH_3Li}{\longrightarrow} \quad CH_3TiCl_3(ether)_2 \atop 6$$

Methyltrichlorotitanium (5) can be distilled at ~37 °C/1 torr [30]. The pure crystals (melting point ~29 °C) have a purple color and can be stored at low temperatures for weeks [18]. At room temperature it is thermally stable for several hours [18]. Gaseous 5 is stable in the dark [31]. The rate of decomposition of crystalline 5 depends upon the purity of the sample, $TiCl_3$, methane as well as an oily residue being formed [18]. In CH_2Cl_2, 5 forms yellow solutions which are stable for days in the refrigerator [21, 29]. Decomposition in hydrocarbon solvents has been studied [23, 30, 31, 32].

An effective way to enhance the kinetic stability of CH_3TiCl_3 (5) is the formation of octahedral complexes, e.g. with bidentate ligands [33, 34]. For example, according to Thiele a slurry of CH_3MgCl in hexane can be reacted with TiCl$_4$ and the solution treated with 2,2'-bipyridyl to form an almost quantitative yield of the adduct 7 [24a]. It is a red-violet, diamagnetic crystalline compound which decomposes at 180 °C. It is also much less air-sensitive than uncomplexed 5 [24a].

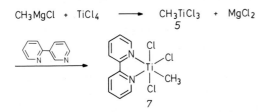

$$CH_3MgCl + TiCl_4 \longrightarrow CH_3TiCl_3 + MgCl_2$$
$$5$$

A number of other crystalline adducts have been isolated (Table 1). It is important to note that 5 also forms bis-adducts with diethylether, but in this case thermal stabilization does not result [33]. In fact, the rate of thermal decomposition is higher than in non-ethereal solutions. Possible

Table 1. Adducts of CH$_3$TiCl$_3$ (5) Obtained in Crystalline Form[a]

Complex	Color	Lit.
5 · (O O ring)	violet	18, 24 b
5 · (S O ring)	red-violet	24 b
5 · 2 N (pyridine)	violet	24 b
5 · 2 CH$_3$CN	violet	24 b
5 · 2 S (ring)	red-brown	24 b
5 · 2 S(CH$_3$)$_2$	red-brown	24 b
5 · (CH$_3$OCH$_2$ / CH$_3$OCH$_2$)	pink-violet	24 b
5 · ((CH$_3$)$_2$NCH$_2$ / (CH$_3$)$_2$NCH$_2$)	violet	36
5 · (Ph$_2$PCH$_2$ / Ph$_2$PCH$_2$)	orange-red	24 b
5 · (CH$_3$O / (CH$_3$)$_2$N ring)	brown-violet	36

[a] For a more complete survey see lit. [33]. A discussion of the structure of similar octahedral complexes as derived from NMR spectroscopy is presented in Section 2.5.

mechanisms of decomposition have been discussed [18, 24, 32, 35]. Nevertheless, useful chemistry can be performed in ether or in THF at temperatures below −20 °C (Chapter 5).

The reaction of pure CH_3TiCl_3 (5) with 18-crown-6-ether in CH_2Cl_2 leads instantly to a deep cherry red solution [37]. Examples of the use of chiral bidentate ligands have also been reported, e.g., 8 [38] and 9 [37] starting from spartein and (−)(R,R)-N,N,N′,N′-tetramethylcyclohexane-1,2-diamine, respectively. All of these complexes can be used for certain C—C bond forming reactions, e.g., addition to aldehydes [38] (Chapter 5).

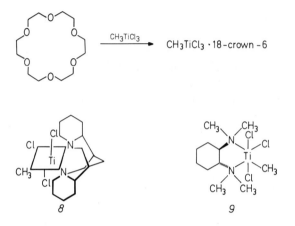

8   9

Whereas definitive structural data have not been obtained in all cases (e.g., the dioxane adduct of 5 is believed to be polymeric, i.e., not to be a simple 1:1 bidentate adduct), a highly interesting X-ray crystallographic study of the diphosphine chelate 10 has been reported by Green [39].

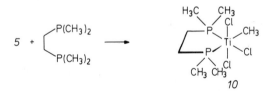

10

The results [39] show that titanium adopts a distorted octahedral geometry. Interestingly, the methyl group displays pronounced distortion in that one of the hydrogen atoms finds itself in unusually close proximity to the titanium atom. Thus, the Ti—C—H angle is 70(2)° and the Ti—H distance amounts to only 2.03 Å! The authors suggest that the C—H group behaves essentially as a lone pair (Scheme 1) which donates electron density to the empty titanium orbitals, agostic hydrogen being involved [39–40]. They regard such interaction as a model for the transition state of α-elimination (1,2-hydrogen shift) and also discuss the mechanism of Ziegler-Natta polymerization. One might expect the phenomenon of agostic hydrogen to be even more important in CH_3TiCl_3 (5) itself, but no X-ray data have been reported to date.

Scheme 1: Schematic Representation of *10* [39]

Generally, CH_3TiCl_3 (*5*) tends to take on two donor moieties to form six-coordinate octahedral complexes as delineated thus far. However, rare cases of penta-coordination have been observed for bulky donor molecules [33], e.g., triphenylphosphine [24b]:

$$PPh_3 + 5 \rightarrow CH_3TiCl_3 \cdot PPh_3$$

Another way to stabilize *5* via coordination is to add ammonium salts [41]. Depending upon the amount of tetraethylammonium chloride (or bromide), anionic compounds of the type *11*, *12*, or *13* are obtained [41]. They all are air- and heat-sensitive, but less so than *5* itself.

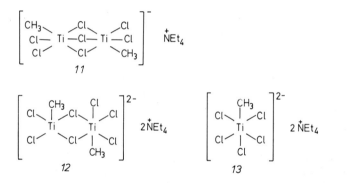

n-Alkyl homologs of $CH_3Ti(OCHMe_2)_3$ (*4*) are thermally less stable than the parent compound [21, 42]. This brings up an important point in the entire organotitanium chemistry. Originally it was believed that C—Ti bonds are thermodynamically weak. However, Lappert showed that this is not the case [43] (see Section 2.2). As pointed out by Wilkinson, the word stable has often been misused and misunderstood, i.e., there has been confusion regarding the terms thermodynamic and kinetic stability [44]. During the last 15 years it has become clear that many transition metal alkyls can decompose via kinetically favored pathways such as β-hydride elimination. Analogous decompositions of non-transition metal alkyls (e.g., RLi or RMgX) require much greater activation energies ΔG^+, i.e., they are kinetically stable. In the case of alkyltitanium compounds, β-hydride elimination is in fact the primary decomposition path [1, 45]. This has been exploited synthetically by Finkbeiner and Cooper [46] and by Asinger [47] in Ti-mediated isomerizations of alkyl Grignard reagents. Catalytic amounts of $TiCl_4$ or other titanium compounds cause magnesium to migrate to terminal positions, e.g.:

23

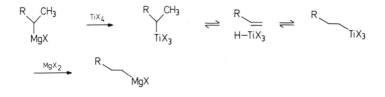

Although the homologs $RTi(OCHMe_2)_3$ generally cannot be distilled without extensive decomposition [48], they are easily handled in solution. For example, ethyl- and n-butyllithium react with chlorotriisopropoxytitanium (3) to form solutions of 14 and 15, respectively, which can be used for further synthetic reactions [21, 49, 50] (Chapters 3 and 5).

$$CH_3CH_2Li \xrightarrow{3} CH_3CH_2Ti(OCHMe_2)_3$$

14

$$CH_3CH_2CH_2CH_2Li \xrightarrow{3} CH_3CH_2CH_2CH_2Ti(OCHMe_2)_3$$

15

The n-alkyl-trichloro derivatives are also considerably less stable than the parent compound CH_3TiCl_3 (5). Solutions are conveniently prepared by reacting dialkylzinc reagents with $TiCl_4$ [24a, 34]. Trichloroethyltitanium (16) has also been prepared by the reaction of tetraethyllead with $TiCl_4$ [18, 25]. Conversion seems to be excellent, but about half of the material decomposes during vacuum distillation. Pure, distilled samples decompose at room temperature within 24 h to form ethane, n-butane and trivalent titanium [25]. In the solid state, compounds $RTiCl_3$ are usually violet-colored; non-ethereal solutions are yellow. All of the titanium species can be stabilized by bipyridyl adduct formation [24a, 34]. This also applies to the allyl derivative, generated by the reaction of allylmagnesium chloride and $TiCl_4$ [51]. Besides zinc, lead or magnesium precursors, organolithium reagents have also been employed, e.g., phenyllithium in the synthesis of trichlorophenyltitanium [24a]. The interaction of ethylaluminum compounds with $TiCl_4$ also affords compounds with C—Ti bonds. In these cases the rate of decomposition has been closely studied [52].

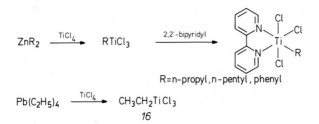

The X-ray structure of the diphosphine adduct *17* of trichloroethyltitanium (*16*) reveals some interesting features [53]. The Ti—C—C angle of 85.9(6)° clearly shows that the methyl group is attracted by the titanium center. One of the hydrogens of this group points toward titanium, the through-space Ti—H distance being only 2.29 Å (which is clearly shorter than the sum of the van der Waals' radii). This effect has been attributed to direct interaction of a C—H moiety with titanium in terms of a two-electron three-centered molecular orbital system [40, 53]. It is also believed to be a model for the transition state of the widely occurring β-hydride elimination of transition metal-alkyls. However, *17* does not readily undergo such decomposition. This decreased tendency can be attributed to a combination of electronic and steric factors. Simply stated, vacant coordination sites in the non-complexed EtTiCl₃ are occupied by the phosphine ligands. The potential role of vacant phosphorus d-orbitals has not been considered. Parenthetically, an ¹H-NMR study of *17* reveals that above 0 °C the four methyl groups and the two methylene moieties become equivalent. On the NMR time scale the ethyl and Cl [1] ligands are thus rapidly interchanging their positions [53].

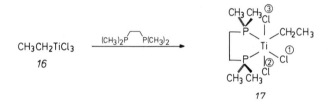

$$CH_3CH_2TiCl_3$$
16

If the alkyl group attached to titanium contains donor heteroatoms, external ligand systems are not necessary to promote stabilization. For example, the red-violet colored 3-(N,N-dimethylamino)-trichlorotitanium (*18*) is "surprisingly" stable, decomposing at 54 °C with formation of ethylene, propene and cyclopropane [54]. Thiele has invoked intramolecular coordination to explain the enhanced thermal stability of the Ti—C σ-bond [54].

$$Zn[(CH_2CH_2CH_2NMe_2)_2]_2 \xrightarrow[-60\,°C]{TiCl_4} Me_2NCH_2CH_2CH_2TiCl_3$$
18

Branched alkyltitanium compounds such as Me₂CHTi(OCHMe₂)₃ or Me₃CTi(OCHMe₂)₃ have not been prepared. They are likely to undergo β-hydride elimination resulting in decomposition or isomerization (with formation of Ti(III)-species), in contrast to the stable zirconium analogs [55]. These comments also apply to branched trichlorides RTiCl₃, which have been postulated as intermediates in the reaction of TiCl₄ with alkylaluminum compounds [56]. An exception is cyclopropyltriisopropoxytitanium, prepared from cyclopropyllithium and chlorotriisopropoxatitanium [21, 50a]. The compound is stable because β-hydride elimination would lead to the highly strained cyclopropene. Trichlorovinyltitanium appears to be a rather sensitive compound which must be handled in solution below $-30\ ^\circ C$ [57a], but highly substituted derivatives can be isolated [57b].

Considerably more stable are certain other derivatives of RTiX$_3$ which lack β-hydrogen atoms, e.g.:

Me$_3$SiCH$_2$Ti(OCHMe$_2$)$_3$ [50a, 58]   Me$_3$SiCH$_2$TiCl$_3$ [59]

Organyltitanium compounds having three N,N-dialkylamino ligands posses unusual thermal stability, as shown by Bürger [63]. Even highly branched alkyl derivatives (isopropyl and *tert*-butyl!) are accessible, which does not apply to the trichloro derivatives due to β-hydride elimination with concomitant formation of Ti(III). An important aspect which has been sometimes overlooked has to do with the possibility of reduction of the metal Ti(IV) → Ti(III) during the synthesis, i.e., prior to the formation of alkyl-titanium bonds. It is not always easy to distinguish between these two modes of Ti(III) formation. In any case, the tendency to form undesired Ti(III)-species in the reaction of organometallics with ClTiX$_3$ increases in the series X = NR$_2$ < OR < Cl. Thus, it is most likely to occur with TiCl$_4$. In fact, under certain conditions even trimethylamine will interact with TiCl$_4$ to form Ti(III)-species [64]. One of the important factors contributing to the stability of the amino compounds appears to be of steric nature [63]. In line with this explanation is the observation that the N,N-diethylamino compounds are thermally more stable than the N,N-dimethylamino analogs. The compounds are, however, oxygen- and moisture-sensitive. Pyrolytic decomposition involves ionic (not radical) intermediates [63]. Table 2 summarizes the decomposition temperatures of various derivatives.

R—Li + XTi(NR$_2'$)$_3$ → RTi(NR$_2'$)$_3$
  (MgX)
  X = Cl, Br

All compounds RTiX$_3$ discussed so far have three identical X groups. This need not be the case, mixed systems being accessible via several routes [1]. One of these involves rapid stoichiometric ligand exchange processes. Fortunately, only one product is usually obtained, depending upon the ratio of starting components [19, 42b, 65], e.g.:

2 CH$_3$TiCl$_3$ + CH$_3$Ti(OCHMe$_2$)$_3$ → 3 CH$_3$Ti(OCHMe$_2$)Cl$_2$
   5   4   19 ($\sim$80%)

Table 2. Thermal Stability of some typical Compounds $RTi(NR'_2)_3$ [63]

Compound	Decomposition temp. (°C)
$CH_3Ti(NMe_2)_3$	80
$CH_3CH_2Ti(NMe_2)_3$	70
$CH_3Ti(NEt_2)_3$	120–130
$n\text{-}C_4H_9Ti(NEt_2)_3$	120–130
$i\text{-}C_3H_7Ti(NEt_2)_3$	120–130
$t\text{-}C_4H_9Ti(NEt_2)_3$	120
$CH_2{=}CHCH_2Ti(NEt_2)_3$	130–135

$$CH_3TiCl_3 + 2\ CH_3Ti(OCHMe_2)_3 \rightarrow 3\ CH_3Ti(OCHMe_2)_2Cl$$
$$\underset{5}{} \qquad \underset{4}{} \qquad \underset{20\ (\sim 80\%)}{}$$

$$4 + CH_3\overset{O}{\overset{\|}{C}}Cl \rightarrow 20$$

Compound *19* is a brownish-violet powder which decomposes at its melting point (60–63 °C). It is readily soluble in CH_2Cl_2 or toluene, but less so in pentane [19, 42b]. *20* can also be prepared by the action of acetylchloride on triisopropoxymethyltitanium (*4*) [65]. This is an interesting observation because it means that *4* selectively transfers an isopropoxy ligand (and not the methyl group) onto acid chlorides. The yellow crystals of *20* (melting point 62–64 °C) can be sublimed in high vacuum [19, 42b]. $TiCl_4$ and $TiBr_4$ are also useful for controlled ligand exchange processes [1, 19, 21, 59c, 66, 67].

A different strategy is to introduce the methyl group (or other moieties) by nucleophilic substitution at the end of a sequence [19, 49a, 65]

$$TiCl_4 + Ti(OCHMe_2)_4 \rightarrow Cl_2Ti(OCHMe_2)_2 \xrightarrow{CH_3Li} (CH_3)_2Ti(OCHMe_2)_2$$
$$\underset{21}{} \qquad\qquad\qquad\qquad \underset{22}{}$$

In another approach, dialkyltitanium(IV) compounds (to be discussed in Section 2.1.3) are reacted with one equivalent of an alcohol. Such alcoholysis leads cleanly to alkanes (e.g., methane) and the corresponding mixed ligand system, as in the following example involving optically active menthol [49a]:

(CH₃)₂Ti(OCHMe₂)₂
22

23

2.1.3 Polyalkyl- and Aryltitanium Compounds

Dialkyldialkoxytitanium compounds can be prepared by reacting two equivalents of an organometallic precursor with dialkoxydichlorotitanium, 22 [19, 49a, 65, 68] and 24 [69] being typical examples (conversion >90%). After fritting off the LiCl and removing the ether, 22 can be crystallized from cold pentane or sublimed at 55 °C/0.1 torr [21]; nevertheless, it is thermally less stable than the monomethyltitanium compound 4.

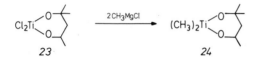

23

24

An equally clean (although less economical) method involves controlled alcoholysis of tetraalkyltitanium compounds (to be discussed later). In an attempt to generate tribenzylethoxytitanium 27 by the addition of only one equivalent of ethanol, the disproportionation products 25 and 26 were obtained [66]. This is an example in which stoichiometric amounts of reagents do not yield a single product. Alcoholysis is particularly useful in case of very sensitive compounds, e.g., 29, as shown by the elegant studies of Jacot-Guillarmod [70].

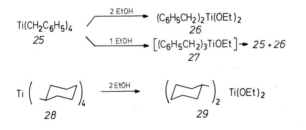

Compounds of the type 29 are thermally unstable and should be handled in solution [70]. Crystals of 24 and 26 have been isolated and X-ray data obtained (Section 2.4); they are actually dimers.

The dichlorides R_2TiCl_2 are also somewhat less stable than the corresponding $RTiCl_3$. They are conveniently prepared by the reaction of $TiCl_4$ with the proper amount of an organometallic species, e.g., CH_3Li [67],

CH_3MgX [12a, 71], $Al(CH_3)_3$ [18, 72] or $Zn(CH_3)_2$ [24a, 73]. The reactions proceed via CH_3TiCl_3 which is methylated by the second equivalent of the organometallic reagent. The reactions using CH_3Li or CH_3MgX proceed in ether and in fact lead to the bis-etherate of $(CH_3)_2TiCl_2$ (*30*) [67]. Ether-free solutions are best prepared by the reaction of $TiCl_4$ with $Zn(CH_3)_2$ in pentane [34, 73] or CH_2Cl_2 [21, 28], conversion being essentially quantitative:

$$Zn(CH_3)_2 \xrightarrow[CH_2Cl_2/-30\,°C]{TiCl_4} (CH_3)_2TiCl_2$$

30

The dichloride *30* has not been characterized as thoroughly as CH_3TiCl_3 (*5*) due to its greater thermal lability [18]. Decomposition pathways have been studied [32]. For synthetic purposes (Chapter 7) it is best handled in solution [28]. The compound forms adducts with dioxane [18a] or tetramethyl ethylene diamine [73], the former being stable at room temperature for days, the latter (characterized by ^{1}H-NMR spectroscopy) slowly decomposing under such conditions.

The reaction of primary or secondary alkylmetal reagents such as $Al(C_2H_5)_3$ or $Al(i-C_4H_9)$ with $TiCl_4$ may lead to R_2TiCl_2, but these compounds have not been characterized due to their extreme thermal lability even at low temperatures ($-60\,°C$) [32, 72, 74]. Reduction to Ti(III) is facile. In contrast, derivatives lacking β-hydrogen atoms can often be isolated, e.g., *31* by distillation at 60–65 °C (1.5 torr) [59]. Whereas dibenzyldibromotitanium *32* is obtained by disproportionation of tetrabenzyltitanium (*25*) with $TiBr_4$, the analogous reaction with $TiCl_4$ fails to afford the dichloro analog [66]. This has not been explained.

$$Me_3SiCH_2MgCl \xrightarrow{TiCl_4} [Me_3SiCH_2]_2TiCl_2$$

31

$$Ti(CH_2C_6H_5)_4 \xrightarrow{TiBr_4} (C_6H_5CH_2)_2TiBr_2$$

25   *32*

Dialkyldiaminotitanium compounds $R_2Ti(NR_2')_2$ are far less stable than the $RTi(NR_2')_3$ analogs. For example, $(CH_3)_2Ti(NEt_2)_2$ decomposes at temperatures above $-30\,°C$ [75].

A few trialkyltitanium compounds of the type R_3TiX have been synthesized. Whereas pure $(CH_3)_3TiX$ having X = Cl [31, 76], I [77], O-t-Bu [78] could not be characterized satisfactorily due to their high sensitivity, more is known concerning benzyl derivatives [66]. For example, *33* and *34* can be obtained in crystalline form. Table 3 reveals the thermal lability of several benzyltitanium compounds as measured by the amount of Ti(III) which is formed as a result of decomposition.

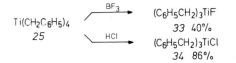

$$Ti(CH_2C_6H_5)_4 \quad \begin{array}{c} \xrightarrow{BF_3} \\ \\ \xrightarrow{HCl} \end{array} \quad \begin{array}{c} (C_6H_5CH_2)_3TiF \\ 33 \quad 40\% \\ (C_6H_5CH_2)_3TiCl \\ 34 \quad 86\% \end{array}$$

Table 3. Percentage of Ti(III) after Thermal Aging at 25 °C for One Day [66]

Compound	% Ti (III)
$Ti(CH_2C_6H_5)_4$	1.7
$Ti(CH_2C_6H_5)_3F$	61.1
$Ti(CH_2C_6H_5)_3Cl$	20.6
$Ti(CH_2C_6H_5)_3Br$	11.5
$Ti(CH_2C_6H_5)_2Br_2$	13.0
$Ti(CH_2C_6H_5)_2(OEt)_2$	1.2

Several groups have studied tetraalkyl- (and aryl)-titanium compounds TiR_4. They are generally referred to as homoleptic σ-hydrocarbyl compounds [45] and are prepared by the reaction of RLi or RMgX with $TiCl_4$ [1a, 45] or in rare cases with $Ti(OC_4H_9)_4$ [79] at low temperatures. Typical examples are *35–40*; their thermal stability varies considerably.

$$Ti(CH_3)_4 \qquad Ti(CH_2CH_2CH_2CH_3)_4 \qquad Ti\left(\text{\raisebox{0pt}{\includegraphics}}\right)_4$$
$$35 \qquad\qquad 36 \qquad\qquad\qquad 28$$

$$Ti(CH_2CH_2CH_2NMe_2)_4 \qquad Ti(CH_2C_6H_5)_4 \qquad Ti(C_6H_5)_4$$
$$37 \qquad\qquad\qquad 25 \qquad\qquad 38$$

$$Ti(CH_2SiMe_3)_4 \qquad Ti(CH_2CMe_3)_4$$
$$39 \qquad\qquad 40$$

$$Ti\left(\text{\raisebox{0pt}{\includegraphics}}\right)_4 \qquad Ti\left(\text{\raisebox{0pt}{\includegraphics}}\right)_4 \qquad Ti\left(CH_2\text{\raisebox{0pt}{\includegraphics}}\right)_4$$
$$41 \qquad\qquad 42 \qquad\qquad 43$$

Tetramethyltitanium (*35*) has no β-hydrogens, but is nevertheless thermally very sensitive, the yellow crystals (free of ether?) decompose at above -25 °C [80]. Decomposition products are methane and a black powder which hydrolyzes to yield more methane as well as small amounts of alkanes.

For synthetic purposes (Chapter 3) the compound can be handled in solution at -30 °C [49c, 81]. In comparing $Ti(CH_3)_4$ with $Si(CH_3)_4$, Lappert concludes that the ease and type of decomposition of the former relative to the latter has to do with the fact that titanium, being a d° transition metal, can readily expand its coordination sphere and make use of low-lying d-orbitals [82]. Tetramethyltitanium *35* (and other TiR_4) can be stabilized by adduct formation with neutral donor ligands such as dioxane, amines, phosphines, pyridine or 2,2′-bipyridyl. However, on heating, such adducts may explode [80].

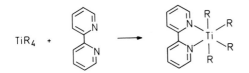

Even more labile and not well characterized are homologs of *35* such as tetraethyl-, tetra-*n*-butyl- (*36*) or tetracyclohexyltitanium (*28*) [79, 83 a, b]. They can be handled only in solution at low temperatures. Attempts to prepare and characterize tetraallyltitanium failed; it may have fleeting existence, decomposition leading to triallyltitanium [83 c]. Tetravinyltitanium has been prepared in solution, but is very unstable [79 a].

One of the first homoleptic σ-alkyl complexes of titanium(IV) to be isolated was tetrabenzyltitanium (*25*) [84, 85]. In the solid state it is stable at 0 °C, in the range 80–100 °C decomposition sets in with formation of toluene, dibenzyl, benzene and other products. The black pyrophoric solid which is also formed reacts with H_2O to afford H_2, methane and ethane. At 25 °C in toluene, decomposition of *25* is slow (1.7 % after 8 h) [66]. The X-ray structure has been published [85] (Section 2.3).

Good yields of tetraphenyltitanium (*38*) are obtained upon reaction of phenylmagnesium bromide with the *bis*-pyridine adduct of $TiCl_4$ or similar complexes $TiCl_4L_2$ in ether [83 b, 86]. In solution, *38* is fairly stable (e.g., in refluxing ether), but in the solid state deterioration is rather rapid at room temperature.

An interesting series of unusually stable, isolable compounds *39* [87], *40* [88], *41* [89], *42* [90] have been prepared and studied in detail [45]: *39* and *40* have no β-hydrogen atoms, *41* and *42* cannot undergo β-hydride elimination due to Bredt's rule. Thus, the half-life of decomposition of *40* at 60 °C is 14 hours [88]. Compound *42* (m.p. 233–235 °C) is also chemically resistant. The reaction with a mixture of HNO_3, HF and H_2O_2 at 170 °C is very slow! The Ti—C bonds in all of these compounds are sterically shielded. In view of these observations, it is surprising that *43* is much less stable; in the solid state it must be stored at or below 10 °C [91].

A novel class of compounds results upon adding alkyllithium reagents to certain tetraalkyltitanium compounds [92]. These ate complexes are stabilized by dioxane or pyridine. For example, the reaction of methyllithium with tetramethyltitanium affords an ate complex which combines with two equivalents of dioxane to form $Ti[(CH_3)_5Li] \cdot 2\,C_4H_8O_2$. This composition is in accord with the elemental analysis. The lemon-yellow crystals appear to be slightly more stable than $Ti(CH_4)_4$ itself. Other derivatives are quite sensitive, some explode on being touched.

$$TiR_4 + R'Li \rightarrow [TiR_4R']Li$$

2. Synthesis and Properties of Some Simple Organotitanium Compounds

Data concerning the precise structure of such adducts is not yet available [93]. An octahedral geometry appears likely. However, higher coordination is also possible, as has been shown to be the case in certain other titanium compounds, e.g., tetrakis-(N,N-diethylhydroxylamido(1-)-O,N)titanium(IV), which is a distorted dodecahedron with a coordination number of eight [94].

The thermal decomposition of several isolable derivatives of the type $[RTi(CH_2C_6H_5)_4]$ Li (R = CH$_3$, C$_2$H$_5$, n-C$_4$H$_9$) has been studied in detail using various analytical tools including ESR measurements [93]. In the temperature range from -30 to 0 °C, Ti(II) species as well as various gaseous products are formed. β-Hydride elimination is the primary process in case of the n-butyllithium adduct. Also, the study shows that Ti(III) ate complexes of the type $[Ti(CH_2C_6H_5)_4]$ Li are not long-lived (as previously purported by other authors), but rather disproportionate to Ti(II) and Ti(IV) compounds [93].

Scheme 2. Thermal Decomposition of an Ate Complex in the Solid State [93]

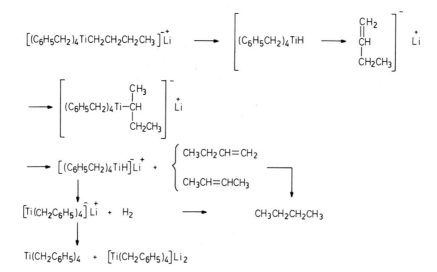

2.1.4 h⁵-Cyclopentadienyltitanium(IV) Compounds

Dozens of Ti(IV) compounds incorporating one or two h^5-cyclopentadienyl (Cp) groups have been prepared [1]. Only a few will be discussed here. One of several synthetic methods involves the reaction of cyclopentadienyl anions (Li or Na salts) with TiCl$_4$ to form crystalline *43* [95] or *44* [96], depending upon the ratio of components used. The organic ligand is always pentahapto-π-bonded. A rare case of σ-bonding is observed in tetra-cyclopentadienyltitanium (*45*) [97].

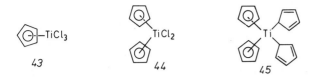

43   44   45

The Cp-group has a strong electron-donating effect. Compound *43* is thermally much more stable than the σ-bonded compound CH_3TiCl_3 (*5*). Furthermore, *43* is a considerably weaker Lewis acid than $TiCl_4$. Thus, NMR studies reveal little or no interaction with THF. However, a few Lewis acid/Lewis base adducts with such powerful bidentate ligands as 2,2′-bipyridyl or o-phenylene-bis(dimethylarsine) are known [95b, 98]. The Lewis acidity of the bis(h^5-cyclopentadienyl)titanium compound *44* is even less pronounced [1].

The chlorine moieties in *43* and *44* can be substituted by a variety of other ligands, including alkyl or aryl groups [1]. It is generally accepted that h^5-cyclopentadienyl ligands have a "stabilizing" effect on the Ti-alkyl bond and that this is due to the occupation of coordination sites which would otherwise be involved in decomposition processes. For example, thermal stability increases in the series $Ti(CH_3)_4 < CpTi(CH_3)_3 < Cp_2Ti(CH_3)_2$. Also, Lewis acidity decreases drastically in this series. Whereas $Ti(CH_3)_4$ forms many adducts with Lewis bases (Section 2.1.3), $CpTi(CH_3)_3$ fails to afford similar compounds, e.g., with pyridine or 2,2′-bipyridly [99].

A convenient synthetic procedure involves treatment of proper chloro-titanium precursors with alkyllithium reagents.

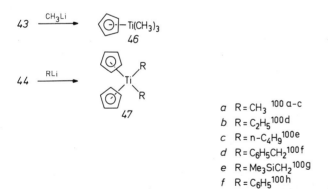

a   R = CH$_3$ 100 a-c
b   R = C$_2$H$_5$ 100 d
c   R = n-C$_4$H$_9$ 100 e
d   R = C$_6$H$_5$CH$_2$ 100 f
e   R = Me$_3$SiCH$_2$ 100 g
f   R = C$_6$H$_5$ 100 h

The thermal decomposition of the moderately stable dimethyl derivative *47a* (m.p. 97 °C with dec.) has been studied in detail [100a–c]. The air-stable orange-yellow compound is fairly unreactive towards cold H_2O. The ethyl and n-butyl analogs are thermally less stable, e.g., *47b* decomposes slowly at room temperature and *47c* deteriorates at −50 °C [100d–e]. As expected, derivatives *47d–f* and related compounds possess considerably greater stability [100f–h].

2. Synthesis and Properties of Some Simple Organotitanium Compounds

The stabilizing effect of h⁵-cyclopentadienyl groups has an important bearing on the use of this ligand in adjusting carbanion selectivity and reactivity via titanation (Section 1.1). Thus, replacing a chlorine or alkoxy ligand in compounds of the type $RTiCl_3$ or $RTi(OR')_3$ by Cp-groups reduces reactivity considerably (Chapters 3-5). Of all ligands at titanium studied thus far, the h⁵-cyclopentadienyl group exerts perhaps the most pronounced electronic and steric effect (Section 2.5.3).

Unsymmetrically substituted compounds *48* or *49* are also accessible [101], as are derivatives with substituted Cp-ligands [102].

$$44 \xrightarrow{R^1Li} Cp_2Ti \!\!\begin{array}{c} R^1 \\ \diagdown Cl \end{array} \xrightarrow{R^2Li} Cp_2Ti \!\!\begin{array}{c} R^1 \\ \diagdown R^2 \end{array}$$

$$\qquad\qquad\quad 48 \qquad\qquad\qquad\quad 49$$

$$R^1, R^2 = alkyl, aryl$$

The chloride *48* readily reacts with CO to form acyltitanium compounds of the type *50*, as shown by Floriani [103]. In case of *47* or *49*, dicarbonyl-*bis*(cyclopentadienyl)titanium (*51*) is formed [103, 104]. The latter smoothly reacts with certain alkyl iodides to form *52*, which are iodo analogs of *50* [105].

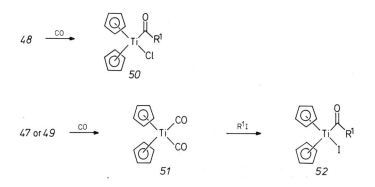

A number of titanacycles have been synthesized using various routes [1c]. Their thermal stability varies considerably. For example, *53* does not decompose at 300 °C (several hours) [106], while *54* has a half life of $t_{1/2} = 0.5$ h at 0 °C in $CFCl_2CClF_2$, yielding primarily ethylene and 1-butene [100e, 107]. Labelling experiments point to an equilibrium between *54* and the *bis*(ethylene) complex *55* [108], a process which is orbital symmetry allowed [109]. Carbonylation at −55 °C affords a new titanium compound *56*, which decomposes at higher temperatures to afford cyclopentanone in good yield [100e]. Titanacyclobutanes [110, 111] are important compounds in olefin metathesis and other processes and will be discussed in Chapter 8.

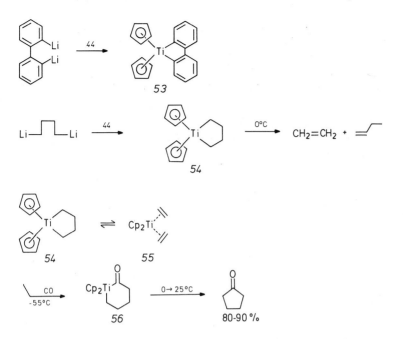

A synthetically important *bis*(cyclopentadienyl)titanium compound is the Tebbe reagent *57* [112]. It is useful in Wittig-type olefinations [113] as well as in the synthesis of metallocyclobutanes, as shown by the recent work of Grubbs [111] (Chapter 8).

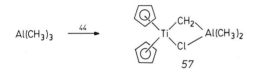

Hydridic compounds of the type *58a* are incapable of (longlived) existence, in contrast to the zirconium analog *58b* (Schwarz reagent) which has considerable synthetic organic utility [114]. Hydride reduction of *44* actually leads to the dimeric Ti(III) compound [Cp₂TiCl]₂ [115].

58   *a* M = Ti
            b M = Zr

Titanium hydrides are known in case of certain Ti(III) compounds (Cp₂TiH); they are dimeric with hydrogen bridging [116]. Syntheses and reactions of other interesting cyclopentadienyltitanium compounds have been reviewed elsewhere [1]. Finally, it is worthwhile mentioning that *bis*(cyclo-pentadienyl)titanium (i.e., titanocene) *59a* has not been prepared due to its high instability, despite many approaches [1]. For this reason, the decamethyl derivative *59b* synthesized by Bercaw, deserves particular attention [117a]. *Bis*-benzenetitanium *60* was first synthesized by Green [118]. Another interesting sandwich compound is Wilke's *bis*(COT)titanium (*61*) [117b]. It undergoes a dynamic process in which the COT rings alternate in bending and flattening [117c].

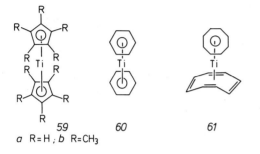

59
a R=H; b R=CH₃

60

61

2.2 Bond Energies

For a long time it was thought that transition metal to carbon σ-bonds are intrinsically weaker than bonds between carbon and non-transition elements, and that metal alkyls must be diamagnetic or coordinatively saturated (18 electron rule) to be "stable". A series of papers by Wilkinson [44] and Lappert [45] dispelled this fallacy (see also Section 2.1.2). Homolytic breakage of transition metal M—C bonds depends on the M—C bond strength, and such processes are actually not as common as previously believed. The first thermodynamic evidence that Ti—C are *not* unduly weak was published by Tel'noi and coworkers [119]. They measured the heats of combustion of *62* and *63*. From this data the bond dissociation energies of Ti—CH₃ and Ti—C₆H₅ were estimated to be 250 kJ/mol and 350 kJ/mol, respectively. Recently, para-substituted derivatives of *63* have been studied. It was found that such substituents have little effect upon the titanium-aryl bond strength [120a].

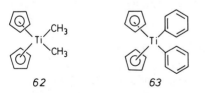

62

63

Definitive thermochemical data was obtained by Lappert for a variety of titanium, zirconium and hafnium compounds [43]. The heats of alcoholysis in isopropanol of several MR_4 compounds, of $M(NR_2')_4$ and of MCl_4 as well as the heats of solution in isopropanol of $M(OPr^i)_4$, RH, R'_2NH and HCl were measured. From these and subsidiary data, standard heats of formation and thermochemical mean bond energy terms $\bar{E}(M-X)$ were derived (Table 4).

Table 4. Thermochemical Data (kcal/mol) for Some Compounds MX_4 [43]

Compound	ΔH_{obs}	ΔH_{vap}	ΔH_i^0 [a]	M—X	$\bar{E}(M-X)$ [b]
$TiCl_4$	-52.5 ± 1.2	9.8^c	-192.2 ± 1.0^c	Ti—Cl	102.7^c
$Ti(CH_2SiMe_3)_4$	-151.8 ± 1.4	18	-205.9 ± 1.5	Ti—C	64
$Ti(CH_2CMe_3)_4$	-203 ± 1.8	21	-58.6 ± 1.9	Ti—C	44
$Ti(CH_2Ph)_4$	-164.4 ± 1.6	21	$+77.6 \pm 1.7$	Ti—C	63
$Ti(NMe_2)_4$	-78.0 ± 0.9	14	-77.3 ± 1.3	Ti—N	81
$Ti(NEt_2)_4$	-80.6 ± 0.7	16^d	-132 ± 1	Ti—N	81^d
$Ti(OPr^i)_4$	-16.2 ± 0.2	17^d	-390 ± 2	Ti—O	115^d
$ZrCl_4$ (c)	-41.7 ± 1.6	26.3^c	-234.35 ± 0.1^c	Zr—Cl	117.4^c
$Zr(CH_2SiMe_3)_4$	-173.3 ± 4.8	18	-215.6 ± 4.8	Zr—C	75
$Zr(CH_2CMe_3)_4$	-230.2 ± 3.8	21	-62.6 ± 3.8	Zr—C	54
$Zr(CH_2Ph)_4$	-181.4 ± 0.1	21	$+63.4 \pm 0.6$	Zr—C	74
$Zr(NMe_2)_4$	-98.7 ± 0.9	17	-87.8 ± 1.3	Zr—N	91
$Zr(NEt_2)_4$	-112.3 ± 0.1	16	-131.3 ± 0.7	Zr—N	89
$Zr(OPr^i)_4$	-7.5 ± 0.3	20	-430 ± 2.0	Zr—O	132
$HfCl_4$	-53.2 ± 1.7	25.3^c	-236.7 ± 1.0^c	Hf—Cl	118.9^c
$Hf(CH_2Me_3)_4$	-230.1 ± 5.6	21	-75.6 ± 5.6	Hf—C	58
$Hf(NEt_2)_4$	-104.1 ± 5.2	16	-151.4 ± 5.2	Hf—N	95
$Hf(OPr^i)_4$	-8.0 ± 0.5	20	-442.5 ± 1.8	Hf—O	137

Notes adabted from lit. [43]: a) The error on ΔH_f^0 (g) is assumed to be ± 8 kcal/mol (± 33.5 kJ/mol); b) ± 2 kcal/mol (8.4 kJ/mol); an alternative way of describing bond strengths is in terms of the mean bond dissociation energy $\bar{D}(M-X)$ when, e.g., $\bar{D}(Ti-C_{neopentyl}) = 50\,(209)$, $\bar{D}(Ti-C_{benzyl}) = 54\,(226)$, $\bar{D}(Ti-NMe_2) = 77\,(322)$, $\bar{D}(Ti-O) = 110\,(460)$ kcal/mol (kJ/mol); the difference between the neopentyl and benzyl systems is disguised in the $\bar{D}$ procedure by the considerably greater stability of the benzyl compared with the neopentyl radical; c) "Selected Values of Chemical Thermodynamic Properties", Nat. Bur. Stand. Techn. Note 270, US Government Printing Office, Washington, D.C.; d) Using different calorimetric reactions, mean bond energies $D(Ti-N)$ and $D(Ti-O)$ have been estimated as 73 (305) and 103 (431) kcal/mol (kJ/mol), respectively [D. C. Bradley and M. J. Hillyer, Trans. Faraday Soc. **62**, 2374 (1966)].

Table 4 shows that the $\bar{E}(M-X)$ values for Ti, Zr and Hf decrease in the sequence $M-O > M-Cl > M-N > M-C$, and that they are monotonically higher as the mass of M increases. Surprisingly, the Ti—C bond in $Ti(CH_2CMe_3)_4$ is considerably weaker than in $Ti(CH_2SiMe_3)_4$. This has been atributed to a substantial steric effect [43]. It should be noted that the E(M—C) values differ from the so called mean bond dissociation energies $\bar{D}(M-C)$ which have been used in other cases (see footnote b of Table 4). In any case, it is clear that the Ti—C bond is not weaker than the M—C bond involving most main group metals [120b]. The other pertinent point to be noted is the pronounced strength of the Ti—O bond. Thus, reactions leading to such bonds are expected to have a strong driving force (Chapters 3–8).

The difference in kinetic stability of main group and titanium σ-alkyls thus relates to the greater readiness of titanium to expand its coordination sphere and/or to provide low lying d-orbitals in transition states [43–45]. β-Hydride elimination is an important, but not the only decomposition pathway $64 \rightarrow 65 + 66$. For example, bi-nuclear processes have been postulated for the decay of tetraalkyltitanium compounds according to $67 \rightarrow 68 \rightarrow 69 \rightarrow 70$ [45a, 121]. In case of Ti(CH$_3$)$_4$ (35), an alternative mechanism, also non-radical in nature, involves 1,2-elimination $35 \rightarrow 71 + 72$ [122]. It is important to remember that in some cases impurities catalyze decomposition, and that autocatalysis may also be responsible for low kinetic stability [80].

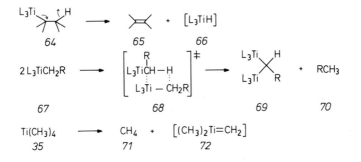

2.3 Bond Angles and Lengths

X-Ray crystallographic and (in a few cases) electron diffraction studies show that monomeric titanium(IV) compounds are tetrahedral, although distortions often occur [1]. Ideal tetrahedral geometry has been observed for TiCl$_4$ (TiCl bond length = 2.17 Å), as demonstrated by electron diffraction [123] and X-ray crystallography [124]. Table 5 displays some typical lengths for Ti—C and Ti—O bonds and includes those of a few other common metal systems.

Bond distances (Table 5), particularly those of metal-oxygen bonds, are rather important in understanding certain stereoselective reactions (Chapter 5). The Ti—O bond ($\sim$1.75 Å), is fairly short relative to Zr—O, Li—O or Mg—O analogs. It should be noted that the Ti—O value may vary considerably (up to 2.1 Å), depending upon the particular type of Ti(IV) compound. For example, it is not surprising that the large values are observed in case of Lewis acid/Lewis base adducts (six-coordinate octahedral species) or dimeric structures formed via Ti—O—Ti bridges (Section 2.4).

Table 5. Typical Bond Lengths

Metal	Metal—Carbon Bond Length (Å)	Metal—Oxygen Bond Length (Å)
Ti	~2.10	1.70–1.90
Zr	~2.20	2.10–2.15
Li	~2.00	1.90–2.00
Mg	~2.00	2.00–2.13
B	1.5–1.6	1.36–1.48

The crystal structure determination of tetrabenzyltitanium (*25*) at −40 °C and at room temperature shows clear deviations from ideal tetrahedral geometry at titanium and at the benzyl C-atoms (Fig. 1) [85]. The phenyl rings are oriented in such a way that their π-faces tilt toward the Ti-atom. This has been interpreted as an electronic interaction between the π-cloud and the empty *d*-orbitals at titanium [125]. Since tetraalkyltitanium compounds are effective Lewis acids (Section 2.1.3), the above effect can be viewed as an intramolecular Lewis acid/Lewis base interaction. Similar distortions have been found for tetrabenzylzirconium [126], but not for tetrabenzylstannane [127]. The Lewis acidity of TiR$_4$ plays an important role in certain stereoselective addition reactions to cyclic phenylketones in which intermolecular complexation between the titanium reagent and the π-face of the aromatic ring exerts a directive effect (Chapter 5).

Distortion from ideal tetrahedral geometry is also observed in the (thermally rather stable) dibenzylderivative *73*, the reason being the different bulkiness of the ligands [128]. The M—Ti—N and C—Ti—C angles are 120° and 99°, respectively. The Ti—N bond length (1.92 Å) is normal. The Ti—C distance (2.09 Å) is comparable to that in tetrabenzyltitanium (2.13 Å). It has been said that this supports the contention that the difference in thermal stability of *25* and *73* is not due to different Ti—C bonds; rather, different modes of decomposition are involved [128]. Also, in *73* there is no Ti-phenyl throughspace interaction, very likely due to steric reasons.

73

As far as donor complexes of Lewis acidic titanium compounds is concerned, the crystal structures of octahedral diphosphine adducts of CH$_3$TiCl$_3$ and C$_2$H$_5$TiCl$_3$ have already been discussed in Section 2.1.2. TiCl$_4$, which has very similar Lewis acidic properties, forms complexes

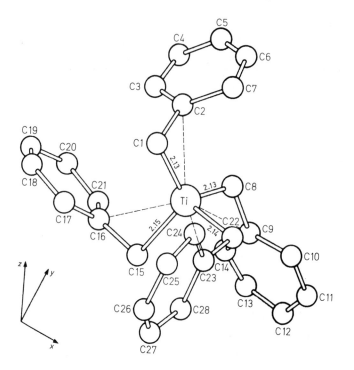

Fig. 1. The Crystal Structure of Ti(CH$_2$C$_6$H$_5$)$_4$ [reproduced with permission from J. W. Bassi, G. Allegra, R. Scordamaglia and G. Chioccola, J. Am. Chem. Soc. *93*, 3787 (1971]

with many donor molecules, such as ethers, ketones, aldehydes and esters [129]. In fact, TiCl$_4$, has been used as an NMR shift reagent [130]. Strangely enough, the adducts of ketones and aldehydes have not been studied by X-ray crystallography, inspite of the fact that many stereoselective C—C bond forming reactions involve such TiCl$_4$-activated forms of carbonyl compounds (Chapter 5). Definitive structural work in this area is badly needed.

In contrast to ketones and aldehydes, related X-ray crystallographic data of several TiCl$_4$-ester adducts are available. TiCl$_4$ reacts with ethyl acetate to form three different products, depending upon the relative amount of reagents used [131]: TiCl$_4$ · 2 CH$_3$CO$_2$C$_2$H$_5$, TiCl$_4$ · CH$_3$CO$_2$C$_2$H$_5$ and 2 TiCl$_4$ · 2 CH$_3$CO$_2$C$_2$H$_5$. The structure of the latter has been elucidated by X-ray crystallography [132]. The yellow, hygroscopic crystals (melting point 102 °C) can be sublimed. They are monoclinic, having a space group P2$_1$/a with four formula units of TiCl$_4$ · CH$_3$CO$_2$C$_2$H$_5$. The dimeric compound (Fig. 2) has two chlorine bridges holding the two titanium atoms together [132]. Titanium is thus octahedrally coordinated by five chlorine atoms and the carbonyl oxygen atom of ethyl ester. The Ti—O distances in the ethyl ester adduct are about 2.03 Å. The Ti—Cl bonds involving the non-bridging chlorines (~ 2.22 Å) are slightly elongated with respect to those in non-complexed TiCl$_4$ (2.17 Å). The Ti—Cl bond length in the bridges are longer (2.5 Å). Complexation also causes a slight increase in the length of the carbonyl C=O double bond [132]. Similar chlorine bridging has been found in [TiCl$_4$ · POCl$_3$]$_2$ [133a] and in the TiCl$_4$ adduct of ethylanisate [133b].

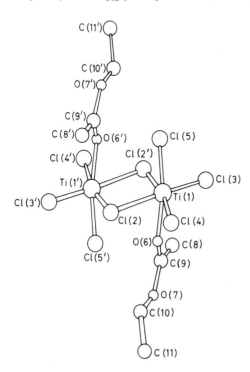

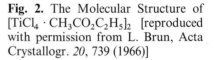

Fig. 2. The Molecular Structure of [TiCl$_4$ · CH$_3$CO$_2$C$_2$H$_5$]$_2$ [reproduced with permission from L. Brun, Acta Crystallogr. *20*, 739 (1966)]

Figure 2 reveals some additional interesting features. Titanium is σ-bonded to the carbonyl oxygen, and not to the ethoxy group. Also, this attachment occurs syn to the methyl group (see partial structure *74*), in contrast to the alternative possibility *75*. A similar phenomenon has been observed for the 2:2 complex of TiCl$_4$ and ethyl-*p*-anisate. The question of syn or anti complexation in case of aldehydes is of great importance in stereoselective addition reactions (Chapter 5).

The first X-ray crystallographic study of a chiral TiCl$_4$-ester adduct (used synthetically for stereoselective Diels-Alder reactions) has been reported by Helmchen [134]. The lactic acid ester derivative *76* was reacted with TiCl$_4$, and crystals of the product *77* were X-rayed. The results (Fig. 3) show some remarkable festures. Firstly, a seven-membered chelate is involved. Also, the enone-moiety does not have an anti-planar conformation as expected, but a syn-planar conformation. Finally, titanium is not in the plane of the ester functions, i.e., they are partially π-coordinated. The Ti—O bond distances are 2.1 Å. Assuming that this compound is the reacting species in solution, the stereoselectivity of the Diels-Alder reaction with cyclopentadiene is explained by the shielding effect of one of the chlorine atoms (the Re-face of the enone-group is sterically inaccessible [134]).

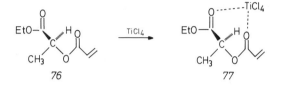

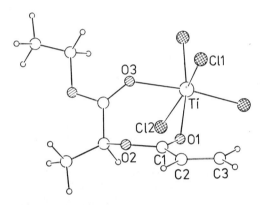

Fig. 3. Crystal Structure of *77* [reproduced with permission from T. Poll, J. O. Metter and G. Helmchen, Angew. Chem. *97*, 116 (1985); Angew. Chem., Int. Ed. Engl. *24*, 112 (1985)]

A number of cyclopentadienyltitanium compounds have also been studied by X-ray crystallography. Prominent examples are CpTiCl₃ (*43*) [135] and Cp₂TiCl₂ (*44*) [136]. The former has the so-called piano stool geometry in which there is some degree of distortion from ideal tetrahedral geometry (Fig. 4). Cp₂TiCl₂ (*44*) also has a distorted tetrahedral arrangement about titanium (Fig. 5). It is not surprising that the Cp—Ti—Cp angle (131°) is larger than the ideal tetrahedral value. In case of the sterically hindered deca-methyl derivative of *44*, several of the methyl groups are forced out of the cyclopentadienyl plane away from the Ti-atom [137].

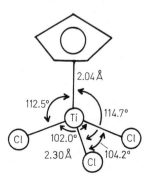

Fig. 4. Geometric Parameters of CpTiCl₃ (*43*) [adabted from P. Ganis and D. Allegra, Atti Accad. Nazl. Lincei Rend. Classe Sci. Mat. Nat. *33*, 303 (1962)]

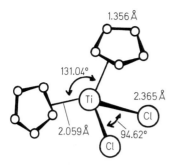

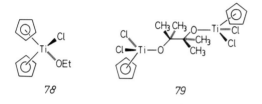

Fig. 5. Geometric Parameters of Cp_2TiCl_2 (*44*) [adapted from A. Clearfield, D. K. Warner, C. H. Salderriaga-Molina, R. Ropal and J. Bernal, Can. J. Chem. *53*, 1622 (1975)]

Caulton has determined the crystal structures of *78* and *79* and was able to make important conclusions regarding the relative electronic effects of Cp, chlorine and alkoxy ligands, respectively [138].

78   *79*

A comparison of *78* with Cp_2TiCl_2 (*44*) shows essential equivalence in C—C and Ti—C distances. However, in *78* the Ti—Cl bond distance is longer by 0.041 Å. Caulton concludes that alkoxides are better π-donors than Cl, i.e., alkoxy groups compete for bonding to an apparently unsaturated metal center more effectively than do chlorine ligands [138]. Ti—Cl lengthening is a manifestation of ethoxide π-donation. The results imply that in Cp_2TiCl_2 (*44*) there is π-donation by chlorine, although very much less than by ethoxide, i.e., "it is not appropriate to think of Cp_2TiCl_2 as an unsaturated (16-electron) complex" [138].

Structural studies of related CpCp'TiCl(OAr) compounds reveal some trends which support the above conclusions [139]. Since the aryloxy groups have ortho substituents, steric inhibition of alkoxide π-donation causes the Ti—O—C angles to range from 140° to 151°, in contrast to the 133° observed for *78*. This is turn results in shorter Ti—Cl distances (2.374 Å) than in *78* (Ti—Cl = 2.405 Å); they are, however, longer than in Cp_2TiCl_2 itself (2.364 Å). Also, steric inhibition results in a longer Ti—O bond in CpCp'TiCl(OAr) (1.88 Å) than that in *78* (Ti—O = 1.855 Å) [139].

An interesting feature of *79* is the Ti—O bond. Its distance (1.750 Å) is 0.105 Å shorter than in *78* and 0.022 Å shorter than in $[CpTiCl_2]_2O$. More remarkable is the greater Ti—Cl bond contraction in going from *78* (Ti—Cl = 2.405 Å) to *79* (Ti—Cl = 2.271 Å). This is explained by the fact that the Ti—Cl bond is mostly σ in character, in contrast to the Ti—O bond with substantial π character [139]. Thus, going from *78* to *79* "demands π-donation from either a σ-bonded chlorine or an already multiply bonded alkoxide". In summary, the OEt and Cl ligands in *78* bond to the formal

It must be remembered that the degree of aggregation of many n-alkoxy and some sec-alkoxytitanium compounds in solution depends upon concentration and temperature [146]. In some cases solvent effects (e.g., benzene vs. dioxane) have been noted [153]. In the concentration range (in benzene) of 0.02 to 0.12 M, tetraisopropoxytitanium (89) shows some degree of association (average association X = 1.4) [145]. Its ^{1}H-NMR spectrum is temperature and concentration dependent [152]. The exchange of bridging and terminal isopropoxy is very fast, since no splitting is observed even below $-50\ ^\circ$C [152]. More bulky secondary analogs are essentially monomeric in boiling benzene [145, 146, 152].

The Trouton constant of the ethoxide (88) is considerably higher than that of the analogous n-propoxide, n-butoxide, n-amyloxide and n-hexyl-oxide [146, 154]. This has been ascribed to the high degree of association of 88 (trimer). Therefore, the Trouton constant is least for 89 (largely monomeric).

Holloway's low temperature ^{13}C-NMR study of various $Ti(OR)_4$ provides additional insight [152b]. Accordingly, straight-chain titanium tetra-alkoxides occur as trimers (Fig. 6), which in turn form higher aggregates at low temperatures. In case of branched-chain derivatives, equilibria between monomeric, dimeric and trimeric forms were proposed for the isobutoxide, and monomer-dimer equilibria for the isopropoxide. Earlier ^{1}H-NMR data [152a] were re-analyzed in order to estimate the enthalpy and entropy of dimerization of the isopropoxide (89): ~ -63 kJ/mol and ~ -226 J K^{-1} mol^{-1}, respectively [152b]. In summary, it is clear that tetraalkoxy-titanium compounds will aggregate if sterically possible, the extent of which depends upon solvent, concentration and temperature. Compounds having groups larger than isopropoxy are monomeric in all physical states and at all concentrations.

Upon replacing alkoxy with chlorine ligands, the aggregation behavior may change [146]. The crystal structure of dichlorodiphenoxytitanium reveals that the compound is dimeric [155a] (Fig. 7). Bridging occurs via oxygen, titanium being pentacoordinate with a trigonal bipyramidal arrangement. The Ti—Cl distances are 2.219 and 1.209 Å, which means slight elongation with respect to Ti—Cl in $TiCl_4$ (2.18 Å). The Ti—O bond lengths are 1.744, 1.910 and 2.122 Å. The shortest of the three involves the bond to the non-bridged oxygen. In benzene the compound is monomeric [155b].

The effect of concentration on the degree of aggregation has been studied for $Ti(O-n-C_4H_9)_4$ and for a series of chlorine containing n-but-oxides [150b]. $Ti(O-n-C_4H_9)_4$ shows distinct concentration dependency; at high concentrations the trimeric forms pertains (Fig. 8). In contrast, no concentration effects are observed for trimeric $ClTi(O-n-C_4H_9)_3$, dimeric $Cl_2Ti(O-n-C_4H_9)_2$ or monomeric $Cl_3Ti(O-n-C_4H_9)$ as illustrated in Fig. 9. This is believed to be due to the presence of electronegative chlorine atoms, which increase the acceptor properties of titanium and therefore also the strength of the alkoxide bridges [146, 150b]. Thus, they do not tend to dissociate even at very low concentrations. Perhaps the nature of the trichloride is the most surprising aspect of this study. It is monomeric just

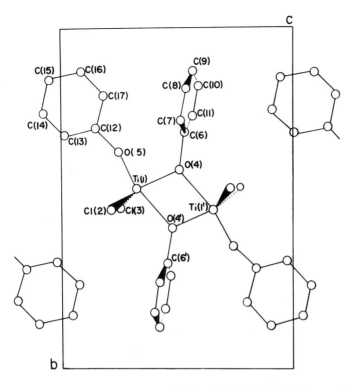

Fig. 7. The Crystal Structure of Dimeric $Cl_2Ti(OC_6H_5)_2$ [reproduced with permission from K. Watenpaugh and C. N. Caughlan, Inorg. Chem. *5*, 1782 (1966)]

like $TiCl_4$ and CH_3TiCl_3, in contrast to dimeric trichlorides such as Cl_3TiN_3 [156].

Turning to alkyltitanium compounds $R'Ti(OR)_3$, the parent member $CH_3Ti(OCHMe_2)_3$ (*4*) has been studied by two groups. According to cryoscopic measurements by Kühlein and Clauss, *4* is slightly associated

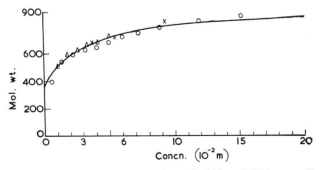

Fig. 8. Variation of Aggregation of $Ti(O-n-C_4H_9)_4$ as a Function of Concentration [reproduced with permission from R. L. Martin and G. Winter, J. Chem. Soc. *1961*, 2947]

Octahedral Ti(IV) compounds are also d°-species, e.g., $Ti(CH_3)_4 \cdot 2$ pyridine (in this case a 12-electron system).

A number of spectroscopic studies of simple organotitanium compounds have appeared [1]. In many cases attempts were made to correlate the data with certain physical and (sometimes) chemical properties. The question of the nature of the titanium-carbon bond (e.g., its polarity) has also been addressed [50a].

2.5.2 Methyltitanium Compounds

This section concentrates on methyltitanium compounds, although comparison with other titanium derivatives are made. The UV spectra of CH_3TiCl_3 (5) and $CH_3Ti(OCHMe_2)_3$ (4), shown in Fig. 11, have been interpreted by Dijkgraaf on the basis of qualitative molecular orbital theory [26]. In case of CH_3TiCl_3, the low intensity ($\varepsilon \sim 75$) of the first absorption at 25 000 cm^{-1} was considered to be a forbidden transition due to the C_{3v} symmetry of the molecules. The second absorption at 43 000 cm^{-1} ($\varepsilon = 14000$) was compared to the 34 840 cm^{-1} band previously observed for $TiCl_4$, which corresponds to a $n \rightarrow \pi^*$ transition. Thus, the second absorption of CH_3TiCl_3 was assigned to an allowed transition of one of the lone electrons of the chlorine atom to a level which is antibonding between titanium and carbon. $CH_3Ti(OCHMe_2)_3$ was analyzed in a similar may [26].

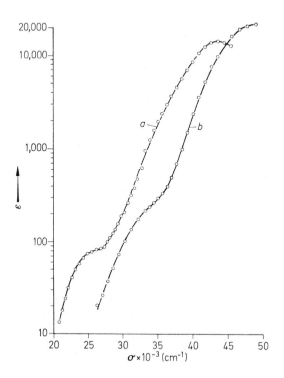

Fig. 11. The UV Spectra of CH_3TiCl_3 (a) and $CH_3Ti(OCHMe_2)_3$ (b) in n-Hexane [reproduced with permission from C. Dijkgraaf and J. P. Rousseau, Spectrochimica Acta *25A*, 1455 (1969)]

In a quantitative study of CH_3TiCl_3 using a modified version of the Wolfsberg-Helmholtz method, MO levels were calculated [164]. The HOMO-LUMO gap turned out to be 3.2 eV. This relative high value was brought in relation to the strength of the Ti—C bond and consequently to the low catalytic effect of CH_3TiCl_3 in Ziegler-Natta polymerization. The charge at titanium as determined by a Mulliken population analysis was calculated to be $+0.54$ (compared to $+0.37$ in $TiCl_4$) [164]. According to a related semi-empirical study, the charges at Ti, C and Cl are $+0.65$, -0.36 and -0.09, respectively, and the HOMO-LUMO difference is 2.99 eV [165].

Armstrong, Perkins and Stewart have published more detailed CNDO-MO-SCF calculations of CH_3TiCl_3 and of other titanium compounds [166]. In case of CH_3TiCl_3 the following charge values were computed: Ti $(+1.23)$, C (-0.37), H $(+0.09)$ and Cl (-0.37). It is difficult to evaluate these and the previously mentioned charge values or to draw quantitative conclusions regarding the polarity of the C—Ti bond, since different assumptions and basis sets were used in the various calculations. Also, it is important to remember the pitfalls in using Mulliken population analyses [167]. Nevertheless, the extreme polarity of the C—Li bond in methyllithium (probably ionic) [168], does not apply to C—Ti bonds [166].

In addition to the charge values, the electronic structure of CH_3TiCl_3 was also calculated by the CNDO-MO-SCF procedure and the results related to spectroscopic and chemical behavior [166]. The computed orbital energies and the correlation diagram derived thereof show that the highest occupied MO relates to the C—Ti σ-bond, the MO's of the chlorine electron pairs lying about 1 eV below. Some degree of Cl_π—Ti_d overlap occurs. The C—Ti anti-bonding σ^*-orbital is almost degenerate, with two sets of π-MO's derived from the 3d orbitals on titanium. In going from CH_3TiCl_3 to $(CH_3)_2TiCl_2$ and $(CH_3)_3TiCl$, the orbital levels shift to higher energies due to the difference in electronegativity of chlorine and carbon [166]. The calculations also show that the HOMO of the dichloride, the monochloride and of $Ti(CH_3)_4$ is again associated with C—Ti σ-bonds. Finally, the UV spectrum of CH_3TiCl_3 was interpreted on the basis of the computational data [166]. In CH_3TiCl_3, the first four transitions are believed to be $\sigma_{C-Ti} \rightarrow d_{Ti}$ with $\sigma_{C-Ti} \rightarrow \sigma^*_{C-Ti}$ being at slightly higher energy. In contrast, the first five transitions in $(CH_3)_2TiCl_2$ are of the type $\sigma_{C-Ti} \rightarrow \sigma^*_{C-T}$. In all of the compounds photo-excitations should cause the σ_{C-Ti} bond to lose an electron which is expected to enter a titanium d-orbital or a σ^*_{C-Ti} orbital. As a result, homolytic rupture of the C—Ti bond is likely to occur, in line with the known photosensitivity of CH_3TiCl_3 and other organyltitanium compounds [166]. Concerning the UV spectrum of CH_3TiCl_3, the calculations predict two weak bands near 3 eV [166], which in fact correlates well with the observed 3.1 eV value [26].

The He(II) photoelectron spectra of CH_3TiCl_3 (5), $CH_3Ti(OCHMe_2)_3$ (4) and several other titanium compounds have been recorded and interpreted with the aid of CNDO/2 calculations [169]. The spectrum of CH_3TiCl_3 (Fig. 12) is very similar to that of $TiCl_4$ [170], (Fig. 13), i.e., the bands at 11.7, 12.7, ∼13.5 and 13.9 eV appear to correspond to the energy levels mainly composed of Cl orbitals. The additional weaker band at 10.8 was assigned to a Ti—C level of a_1 symmetry [169], in agreement with the results of the previous MO study [166].

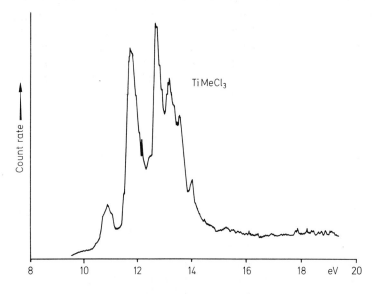

Fig. 12. The Photoelectron Spectrum of CH_3TiCl_3 [reproduced with permission from M. Basso-Bert, P. Cassoux, F. Crasnier, D. Gervais, J.-F. Labarre and P. De Loth, J. Organomet. Chem. *136*, 201 (1977)]

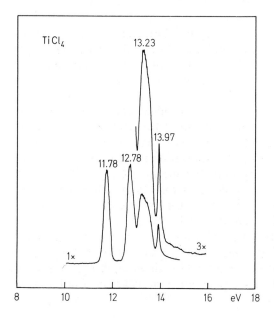

Fig. 13. The Photoelectron Spectrum of $TiCl_4$ [reproduced with permission from P. A. Cox, S. Evans, A. Hamnet and A. F. Orchard, Chem. Phys. Lett. *7*, 414 (1970)]

The computed MO levels (theoretical IP values assuming Koopman's approximation) for CH_3TiCl_3 are listed in Table 6. Although the calculated eigen-values are expected to be lower than the experimental values, the ordering of the levels is likely to be correct. Assignments were proposed as follows: The first band at 10.8 eV (calculated IP = 13.77 eV) corresponds to the $5a_1$ level, representing mainly the Ti—C σ-bond with very little chlorine mixing. The experimental 11.7 eV band corresponds to ionization from the $1a_2$ and 5e molecular orbitals (calculated IP's = 14.7 and 14.65 eV), composed of essentially pure p orbitals of chlorine. The next bands are also pure in Cl p-character or mixed with titanium (Table 6).

Table 6. Experimental and Calculated Ionization Potentials (eV) of CH_3TiCl_3 [169]

Experimental	Symmetry	Calculated	Orbital character
10.8	$5a_1$	13.77	Ti—C
11.7	$1a_2$	14.57	Cl
	5e	14.65	Cl
12.7	4e	15.46	Cl
13.1	$4a_1$	15.75	Cl + Ti
13.5	3e	16.16	Cl + Ti
13.9	$3a_1$	16.35	Cl

In case of $CH_3Ti(OCHMe_2)_3$, the PE spectrum is not as easily analyzed [169]. This is due in part to the overlap of bands. Calculations were performed on the model compounds $CH_3Ti(OCH_3)_3$ and $Ti(OCH_3)_4$. The conclusion was made that the second (9.8 eV) and the third (10.4 eV) experimental bands feature the Ti—C bond with oxygen admixture (orbital character: Ti—C + O). Interestingly, the first band at 9.4 eV has only (Ti + O) bond character.

The amino compound $CH_3Ti(NEt_2)_3$ was also studied and analyzed [169]. Here again there is considerable mixing of the heteroatom (nitrogen) orbitals with those of titanium and carbon (10.1 eV). In case of $CpTiCl_3$ and Cp_2TiCl_2, a decrease ($\sim$1 eV) of the ionization potential of the chlorine orbitals is observed upon replacing one chlorine by a Cp group. This was attributed to the strong π-donor character of the Cp group, which in turn decreases the bonding character of the electron pairs at chlorine [169]. Also, the bands primarily due to the 3p chlorine orbitals are shifted by $\sim$1 eV to lower energy relative to those in $TiCl_4$. This again is evidence for the strong electron donating effect of Cp groups and the decreased π-donation of chlorine. Similar effects are observed for alkoxytitanium derivatives. They are related to the previously discussed phenomena reported by Caulton [138] (Section 2.3).

Turning to nuclear magnetic resonance studies, the ^{1}H-NMR spectrum of CH_3TiCl_3 (5) in relation to those of several other organotitanium compounds have been discussed in terms of bonding. Inspite of the fact that it is not easy to provide simple correlations between chemical shifts and such physical parameters as electronegativity or bond polarity, interesting observations were made. For example, the chemical shift (TMS as standard) of CH_3TiCl_3 (5) is solvent dependent [72, 171] $\delta = 2.78$ in C_2Cl_4 vs. 2.20 in C_6D_6. This has been ascribed to effective solvation of the Lewis acid CH_3TiCl_3 by benzene, and is reminiscent of the charge-transfer interactions of $TiCl_4$ with various aromatic compounds [172]. In case of $(CH_3)_2TiCl_2$ (30) a similar, but smaller effect appears to be operating ($\delta = 2.47$ in C_2Cl_4 and 2.00 in C_6H_6) [171]. It is tempting to relate this phenomenon to the previously discussed intramolecular donor-acceptor interaction in tetrabenzyltitanium (25) [85] (Section 2.3) as well as to the directive effect of phenyl groups in certain stereoselective addition reactions (Chapter 5).

The proton chemical shift of the methyl group in CH_3TiCl_3 (5) has been compared to those of CH_3CCl_3 (2.74), CH_3SiCl_3 (1.44), CH_3GeCl_3 (1.58) and CH_3SnCl_3 (1.65). Although no great NMR spectroscopic correspondence between the group IVa and IVb compounds is to be expected, the values do provide a set of data for compounds of identical substitution and of similar geometry. It becomes clear that the spectrum of CH_3TiCl_3 (5) is almost identical to that of CH_3CCl_3, inspite of the different electronegativities: Ti (1.6) vs. C (2.5). Since Allred-Rochow or Pauling electronegativity values for elements in the lower part of the periodic table have limited meaning [173], various authors have discussed "effective" electronegativity qualitatively and quantitatively with the aid of ^{1}H-NMR data [72, 174]. For example, the effective electronegativity of titanium was calculated to be 1.25 based on chemical shift data of $C_2H_5Ti(NEt_2)_3$ [63a]. Nevertheless, it is difficult to evaluate such numbers.

In other attempted correlations, the position of the ^{1}H-NMR methyl peak of various derivatives was explained by the relative π-donor capacity of the metal ligands [174]. Table 7 shows that for a series of methyltitanium compounds the CH_3-signal appears at lowest field in case of CH_3TiCl_3, and that a large upfield shift is observed upon going to $CH_3Ti(OCHMe_2)_3$ or to $CH_3Ti(NEt_2)_3$. Furthermore, Cp-groups exert a similar effect [174, 175]. The authors conclude that π-donation increases in importance in the order [174]:

$Cl < OR < NR_2 < Cp$.

Although this may in fact be true, the NMR data alone does not provide rigorous proof, since different types of effects in the various compounds may be involved, including geometric changes. A recent ^{1}H-NMR study of $CpTiCH_3(Cl)_2$ has revealed a slight solvent dependency, related to that observed for CH_3TiCl_3 and $(CH_3)_2TiCl_2$; the methyl protons appear at $\delta = 1.93$ (in $CDCl_3$), 1.83 (in CCl_4) and 1.74 (in C_6D_6) [176]. This is not seen in $CH_3Ti(OCHMe_2)_3$: $\delta = 0.57$ (CCl_4) [174] vs. 0.98 (C_6D_6) [20], but such small differences measured on the older instruments may not be meaningful.

Table 7. ^{1}H-NMR Shift Values for some Methyltitanium Compounds (TMS as standard)

Compound	Solvent	CH$_3$-Absorption (δ in ppm)
CH$_3$TiCl$_3$	CCl$_4$	2.70 [174]
CH$_3$TiBr$_3$	CD$_2$Cl$_2$	2.55 [18c]
CH$_3$Ti(OCHMe$_2$)$_3$	CCl$_4$	0.57 [174]
CH$_3$Ti(OCMe$_3$)$_3$	CCl$_4$	0.50 [174]
CH$_3$Ti(NEt$_2$)$_3$	CCl$_4$	0.43 [63a]
CH$_3$Ti(Cl)(OCHMe$_2$)$_2$	CCl$_4$	0.91 [174]
CH$_3$Ti(Cp)(OCHMe$_2$)$_2$	CCl$_4$	0.47 [174]
CH$_3$Ti(Cl)(Cp)$_2$	CDCl$_3$	0.68 [175]

The ^{13}C-NMR (100 MHz) spectra of several methyltitanium compounds have been recorded [158]. The methyl signal of CH$_3$TiCl$_3$ in the decoupled spectrum appears as a sharp singlet at $\delta = 113.9$ (CD$_2$Cl$_2$) or 112.7 (C$_6$D$_6$) at $+33$ °C. Similarly, in case of CH$_3$Ti(NEt$_2$)$_3$ the singlet appears at 30.4 (DCCl$_3$). The spectra of CH$_3$Ti(OCHMe$_2$)$_3$ at various temperatures point to aggregation. At -70 °C (CD$_2$Cl$_2$) the signals of the isopropyl group at 25.7 and 77.1 ppm are broad; also, there is one major CH$_3$Ti-peak at 41.4 ppm and two smaller ones at $\delta = 42.9/49.3$. At $+33$ °C, the isopropyl group gives rise to sharp signals at $\delta = 26.4$ and 76.8, but the CH$_3$Ti-peak cannot be seen due to broadening. At $+81$ °C a small and very broad peak at $\delta = 40$ becomes visible. Other than the aggregation phenomenon, the relative shifts of the three compounds are interesting. They are in line with the electron donating capacity of the heteroatoms in the order N > O $\gg$ Cl. The enormous effect of the trichloro derivative is also seen in trichloro-titanium enolates (Chapter 5).

In principle, ^{47}Ti- and ^{49}Ti-NMR spectroscopy should be a useful tool in studying titanium compounds, the nuclear spins being $I = 5/2$ and $7/2$, respectively [177]. The natural abundance of these two stable isotopes is 7.3% and 5.5%, respectively [177]. Theoretically, methyltitanium compounds could also show proton-titanium coupling. However, nuclear quadrupolar relaxation may render the observation of these phenomena difficult, if not impossible [177]. For example, it was found that a sample of CH$_3$TiCl$_3$ 40% enriched in ^{47}Ti shows no methyl sidebands; quadrupolar relaxation completely decouples the proton spin from the titanium spin [72]. A few cases of successful Ti-NMR spectroscopy are known, however. Examples are the ^{47}Ti- and ^{49}Ti-spectra of TiCl$_4$ and TiBr$_4$ (both of which have T$_d$ symmetry) [178]. It was observed that the magnetogyric ratios of the ^{47}Ti and ^{49}Ti isotopes are quite similar, i.e., the spectra are twinned by 271 ppm. The most significant feature is the fact that the bromide absorbs at lower field than the chloride by ~ 500 ppm, inspite of the fact that the latter is more electronegative. This halogen dependence is opposite to that of MX$_4$ compounds of the main group elements and has been referred to as the "inverse halogen effect". The ^{47,49}Ti chemical shifts were discussed in terms of the parametric contribution, σ_p, to the nuclear magnetic shielding, while the diamagnetic contribution, σ_d, was assumed to remain constant. Since σ_d is inversely proportional to the average excitation energy ΔE of the molecule, a linear relationship between δ (^{47,49}Ti) and $1/\Delta E$ may be anticipated. Experimentally, this is the case [178], i.e., there is a correlation between the Ti-NMR and UV spectra of the compounds.

A relationship between ^{47,49}Ti shift data and the position of the low-lying electronic excited state was also found in a recent, definitive study in which TiX$_4$ (X = Cl, Br, I) and Cp$_2$TiX$_2$ (X = Cl, Br, I) were examined [179]. A few other titanium compounds have also been studied, including Ti(OCHMe$_2$)$_4$; the line width at half-height (78 Hz), is much greater than in case of TiCl$_4$ (3 Hz). This is consistent with the lowering of the symmetry from T$_d$ to (at best) T, which causes a residual electric field gradient at titanium and efficient quadrupolar spin relaxation. The role of a quadrupolar relaxation mechanism was revealed in a detailed study of TiCl$_4$; the quadrupole coupling constants for ^{47}Ti and ^{49}Ti were estimated to be 2.8 and 2.4 MHz, respectively [179]. In another recent study concerning Ti-NMR, it was found that an effect similar to the inverse halogen dependence also operates in going from Cp$_2$TiX$_2$ to the analogous deca-methyl derivatives Cp$_2^*$TiX$_2$, an average downfield shift of ~ 312 ppm being observed [180]. It was concluded that changing from Cp to Cp* has a greater influence on the nuclear property of titanium than does a change in halogen. In Section 2.5.3 the electronic effect of Cp groups is discussed in more detail.

On the basis of Ti-NMR measurements, it was also possible to draw conclusions regarding the aggregation state of liquid TiCl$_4$ [178]: It exists as a monomer, in line with an earlier Raman study [181]. Finally, mixtures of TiCl$_4$ and TiBr$_4$ show just *one* resonance signal, the position of which depends upon the relative amounts of the components. This indicates rapid halogen exchange between the parent compounds and the mixed species TiCl$_3$Br, TiCl$_2$Br$_2$ and TiClBr$_3$, in complete accord with a previous study based on Raman spectroscopy [181]. Unfortunately, methyl derivatives of the type CH$_3$TiX$_3$ or (CH$_3$)$_2$TiX$_2$ have not been studied by ^{47,49}Ti-NMR to date.

Other spectroscopic methods have been applied to methyl compounds CH$_3$TiX$_3$. The IR spectrum of CH$_3$TiCl$_3$ (5) in the solid, liquid and gas phase has been measured several times, but agreement as to all of the assignments has not been reached [182]. The same applies to CD$_3$TiCl$_3$ [182]. The force constants of the C—Ti bond was calculated to be 1.86 N/cm; the values are larger for analogous silicon and tin compounds (CH$_3$SiCl$_3$ (2.9 N/cm) and CH$_3$SnCl$_3$ (2.3 N/cm)) [183].

Recently, a detailed study of the IR and Raman spectra of gaseous and solid CH$_3$TiX$_3$ (X=Cl, Br, I) and CD$_3$TiX$_3$ (X=Cl, Br, I) was published [184]. Torsional barriers to internal C—Ti rotation in the gaseous molecules were evaluated according to the Durig treatment for single-top systems: 6.3 kJ/mol for CH$_3$TiCl$_3$, 5.8 kJ/mol for CH$_3$TiBr$_3$ and 5.4 kJ/mol for CH$_3$TiI$_3$. These values were compared to those of CH$_3$CCl$_3$ (22.6 kJ/mol),

CH_3SiCl_3 (8.8 kJ/mol), and CH_3GeCl_3 (6.1 kJ/mol) [185]. It is clear that secondary overlap and non-bonded interactions decrease as the carbon-metal bond distance increases. However, it is not easy to pinpoint other effects which are certain to be operating in this series (Fig. 14).

M = C, Si, Ge, Ti

Fig. 14. Rotations in Compounds CH_3MCl_3

Staley has studied the mass spectrum of CH_3TiCl_3 as well as its gas-phase ion chemistry using ion cyclotron resonance (ICR) trapping techniques [186]. The 70 eV mass spectrum shows the parent ion $CH_3TiCl_3^+$ and the fragment ions $TiCl_3^+$, $CH_3TiCl_2^+$, $TiCl_2^+$ and $TiCl^+$ having relative abundances of 9%, 100%, 29%, 37% and 15%, respectively. $TiCl_3^+$ reacts with CH_3TiCl_3 via chlorine transfer to give $CH_3TiCl_2^+$ as the major ion at intermediate times. Ethylene was shown to react with $CH_3TiCl_2^+$ to afford $C_3H_5TiCl_2^+$ and H_2. Deuterum labeling studies point to the following insertion mechanism which is believed to be related to the Ziegler-Natta polymerization [186]:

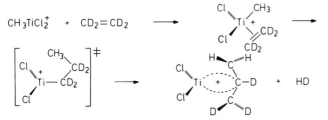

Spectroscopic data of octahedral complexes of CH_3TiCl_3 with donor molecules (THF, $S(CH_3)_2$, TMEDA, etc.) is also available. The ^1H-NMR spectra generally show that Lewis base complexation causes an upfield shift by 0.3–0.5 ppm, relative to CH_3TiCl_3 [24b, 36]. In case of chelation involving bidentate ligands, the meridional form *94* is preferred over the facial diastereomer *95*, as demonstrated by ^1H-NMR spectroscopy [36]. Figure 15 shows the temperature dependency of the adducts *94a–c*.

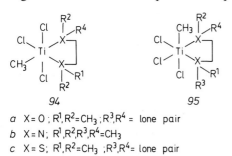

a X = O; $R^1, R^2 = CH_3$; $R^3, R^4 =$ lone pair
b X = N; $R^1, R^2, R^3, R^4 = CH_3$
c X = S; $R^1, R^2 = CH_3$; $R^3, R^4 =$ lone pair

57

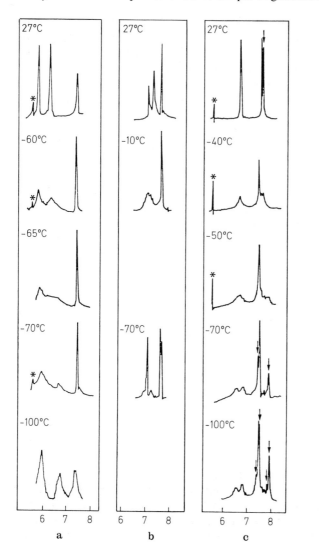

Fig. 15. Variable Temperature ^{1}H-NMR Spectra of Octahedral Complexes *94a–c*
[reproduced with permission from R. J. H. Clark and A. J. McAlees, J. Chem. Soc. A,
2026 (1970)]

The most conclusive evidence for the meridional form originates from
the variable temperature spectra of the CH_3TiCl_3 ($CH_3SCH_2CH_2SCH_3$)
adduct *94c*. Two distinct exchange processes can be frozen out. At 27 °C
three singlets at $\tau = 7.54$, 7.49 and 6.66 are observed, corresponding to the
sulfur-methyl, titanium-methyl and methylene protons. At lower temperatures, the τ 7.54 and 6.66 peaks broaden and resolve to a 1:1 doublet.
This is due to a slowing of the primary exchange process (exchange of R^1,
R^2, R^3, R^4), and not to sulfur inversion. Thus, below -70 °C, further

splitting of each of the sulfur-methyl peaks into an unsymmetrical doublet occurs. It has been claimed that these observations cannot be brought in line with the alternative diastereomer *95 c* [36]. The primary exchange process may be due to rapid opening of the chelate ring and reclosure following free rotation about bonds in the ligands, or to internal twist motion. Finally, it has been noted that sulfur inversion is faster if titanium is coordinated than if such metals as platinum or rhodium are attached. This may be due to the fact that titanium has more available empty *d*-orbitals than platinum(II) for bonding with the sulfur lone electron pair, and, in contrast to the other metals, cannot participate in back-bonding to sulfur [36].

It should be noted that these octahedral complexes bear some relationship to bis(β-diketonato)titanium(IV) compounds, e.g., $Ti(acac)_2X_2$. These compounds are actually (intramolecular) octahedral bis(chelato) complexes which exist as rapidly interconverting stereoisomers [187]. The activation parameters as determined by dynamic NMR spectroscopy are in line with a twist mechanism [187]. It would be interesting to study the dimethyl derivatives ($X=CH_3$):

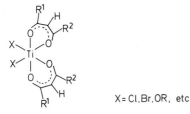

$X = Cl, Br, OR,$ etc.

Turning from mono-methyltitanium compounds to tetra-alkylderivatives TiR_4, several spectroscopic and theoretical investigations deserve attention. The study of the parent compound, $Ti(CH_3)_4$ (*35*) has been hampered by its pronounced thermal instability (Section 2.1.3). In fact, it is not clear whether the pure compound has ever been observed spectroscopically, in contrast to the more stable octahedral donor complexes [80]. The ^{1}H-NMR spectrum in ether (thus, probably and etherate) shows a singlet at unusually high field ($\delta = 0.68$) [80d]. The IR- and Raman spectra were recorded in ether at low temperatures, and on the basis of several assumptions the force constant calculated (2.28 N/cm) [188]. The conclusion that the results are in line with non-complexed tetrahedral $Ti(CH_3)_4$ is interesting, but needs to be checked by other methods.

Several MO calculations of non-complexed $Ti(CH_3)_4$ have been performed. According to a CNDO-MO-SCF study, the computed values of the charge at titanium and carbon are $+0.85$ and -0.39, respectively, compared to $+1.23$ (Ti) and -0.37 (C) calculated by the same method for CH_3TiCl_3 [166]. The *bis*-pyridine adduct $Ti(CH_3)_4 \cdot 2\,C_5H_5N$ was calculated using semi-empirical methods; accordingly, the HOMO-LUMO gap ΔE is only 0.14 eV [165, 189]. As stated previously, it is difficult to assess such calculations. A recent ab initio calculation at the STO-3G level of $Ti(CH_3)_4$ has been performed and discussed by Hehre and co-workers [190]. The computed

2. Synthesis and Properties of Some Simple Organotitanium Compounds

Ti—C bond length of 2.096 Å compares well with that found experimentally for tetrabenzyltitanium (2.170 Å; see Section 2.3). These authors also calculated such species as $H_2Ti=CH_2$, $Cl_2Ti=CH_2$, $H_2Ti=O$, $Cp_2Ti=O$ and titanacyclobutane. It was concluded that some of the current criticism concerning MO calculations of transition metal complexes is well founded (e.g., minimal basis sets do not give reliable orbital energies in such cases), but that many other computational aspects need to be carefully explored before being discarded [190].

The increased thermal stability of other tetraalkyltitanium compounds makes spectroscopic studies easier. For example, the PE-spectra of $Ti(CH_2-SiMe_3)_4$ (*39*) and $Ti(CH_2CMe_3)_4$ (*40*) were recorded and discussed by Lappert, et al. [191]. Table 8 shows the energies of the three highest occupied MO's of *40* as well as those of the Zr, Hf, Ge and Sn analogs. With the exception of the germanium compound, the first IP's are all rather similar. They have been assigned to metal-carbon bonds. Thus, for the isoleptic transition metal compounds, the constancy of the first band indicates a constancy of the central atoms parameters, unless various trends happen to cancel each other. On the basis of the similarity of the three upper PE bands of the transition metal complexes and of the tin compound, similar ground state electronic properties were deduced, which is also consistent with vibrational data [191]. Interestingly, the IP's of $Ti(CH_2SiMe_3)_4$ are slightly higher than those of the neopentyl derivative (*40*), in accord with the higher electron releasing property of the CH_2SiMe_3 ligand.

Table 8. Energies (eV) of the Three Highest Occupied MO's of $M(CH_2CMe_3)_4$ [191]

Metal					Assignment
Ti	Zr	Hf	Ge	Sn	
8.3_3	8.3_3	8.5_1	9.0_1	8.5_8	$\sigma(M—C)$
11.3_5	11.2_8	11.4_0	10.2_8	11.1_6	$(C—C)$
12.5_9	12.5_0	12.5_4	12.2_5	12.3_7	

Recently, the structurally related tetranorbornyl derivative *41* [89] was studied by UV-spectroscopy [192]. The electronic absorption spectrum shows an intense band at λ 245 nm ($\varepsilon = 29\,200$) as well as a very weak band at 367 nm ($\varepsilon = 253$) and shoulders at 312 and 412. The strong absorption was assigned to the allowed ligand-to-metal charge transfer (LMCT) transition $^1A_1(a_1^2t_2^6) \rightarrow {}^1T_2(a_1^2t_2^5e^1)$, which is believed to be responsible for homolytic Ti—C cleavage upon irradiation. Near-ultraviolet photolysis of *41* does in fact generate norbornyl radicals which undergo stabilization reactions to form norbornane and 1,1'-binorbornyl [192]. The molecular orbital diagram for *41* was derived, although the precise ordering of occupied MO's could not be ascertained (Fig. 16).

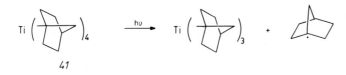

41

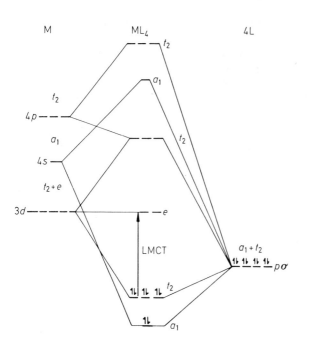

Fig. 16. Qualitative One-Electron MO Diagram for *41* Considering only σ-Bonding [reproduced with permission from H. B. Abrahamson and M. E. Martin, J. Organomet. Chem. *238*, C 58 (1982)]

Relevant to the above discussion is the use of tetraneopentyltitanium (*40*) as a catalyst in the polymerization of styrene and methyl methacrylate under photolytic conditions [193]. The primary process is homolysis of the Ti—C bond, radical polymerization then setting in.

2.5.3 h^5-Cyclopentadienyltitanium Compounds

A number of spectroscopic and theoretical studies of h^5-cyclopentadienyltitanium compounds have appeared [1]. Cp-groups appear as singlets in the aromatic region of the ^{1}H-NMR spectra, the precise chemical shift being very sensitive to the nature of the other three ligands at titanium. The singlet of CpTiCl$_3$ (*43*) is slightly solvent dependent [95]: $\delta = 7.18$ (THF), 7.06 (CDCl$_3$), 7.05 (CH$_2$Cl$_2$), 7.21 (CCl$_4$) and 7.25 (CH$_3$CN). Substitution of chlorine by alkoxy ligands at titanium causes an upfield shift, e.g.; $CpTiCl(OC_2H_5)_2$ $\delta = 6.44$ (in CCl_4) and $CpTi(OC_2H_5)_3$ $\delta = 6.22$ (in CCl_4) [95, 194].

The ^{13}C-NMR spectra of Cp-titanium compounds show a singlet for the five ring C-atoms; here again, the position is sensitive to the nature of the other ligands, e.g., $\delta = 123.1$ ($CpTiCl_3$ in C_6D_6) [194], 114.7 ($CpTiCl(OC_2H_5)_2$ [194] in C_6D_6) and 112.3 ($CpTi(OEt)_3$ in C_6D_6) [194]. Thus, as better π-donors are introduced, the cyclopentadienyl hydrogens and carbons respond by shifting to higher fields. This trend continues upon going to compounds having two Cp-groups. Thus, the ^1H-NMR spectrum of Cp_2TiCl_2 (*44*) shows a singlet at 5.92 (C_6D_6) [195]. In non-aromatic solvents, a shift to lower field is observed: 6.62 (acetone), 6.55 (CH_2Cl_2) and 6.59 ($CDCl_3$) [195].

Nuclear quadrupol resonance spectroscopy (NQR) has been applied to the titanium halides TiX_4, $CpTiX_3$ and Cp_2TiX_2 by Nesmeyanov, who was able to draw conclusions regarding bonding in these compounds [196]. For example, the ^{79}Br- and ^{81}Br-NQR spectra of $TiBr_4$, $CpTiBr_3$ and Cp_2TiBr_2 show that the introduction of Cp groups causes an increase in the nuclear quadrupol coupling constants of bromine. It was proposed that this is in line with the powerful electron-donating property of a Cp-ligand, which causes a weakening of p_π—d_π bonding between titanium and bromine. The same trend was observed in NQR experiments involving ^{35}Cl and ^{127}I. The results were discussed in relation to the known decrease of Lewis acidity in the series $TiCl_4 > CpTiCl_3 > Cp_2TiCl_2$ (Section 2.1.4).

The PE spectrum of Cp_2TiCl_2 was first recorded and interpreted by Dahl, et al. [197]. The molecular orbital energy-level diagram was derived on the basis of non-parameterized Fenske-Hall type MO calculations. The three bands observed between 8 and 9 eV were assigned to the four chlorine p_πMO's, the two strong bands (10.12 eV) to the Cp—Ti interaction through the bonding e_1 orbitals of the ring. Furthermore, the MO calculations suggest that the LUMO of Cp_2TiCl_2 involves the titanium d_{z^2} and $d_{x^2-y^2}$ atomic 3p orbitals of the chlorine ligands. Other interpretations and in part different MO levels have been suggested by various authors [198–203].

A qualitative picture of bonding in Cp_2TiCl_2 and in other transition bis(cyclopentadienyl) metal complexes has been provided by Hoffmann [200]. The MO's of the bent Cp_2Ti-fragment were constructed and interactions with additional σ-ligands considered. Figure 17 shows the MO diagram (in which the electrons of the CpTi fragment have been left off for clarity). Extended Hückel calculations of the model compound Cp_2TiH_2 show that the energy of the non-bonding empty a_1 orbital depends upon the angle $\emptyset$ defined by H—Ti—H [200], which is in line with Dahl's conclusion regarding Cp_2TiCl_2 [197]. The total energy of the hypothetical Cp_2TiH_2, however, does not depend critically on the H—Ti—H bond angle, a fairly shallow minimum occurring at $\emptyset = 110°$.

This picture (Fig. 17) has been adabted by Fay to describe the bonding in the dibenzoate *96* [201]. Extended Hückel calculations on the model compound *97* show that the energy minimum occurs at a distorted tetrahedral arrangement in which the Ti—O—C bond angle $\emptyset$ is 180°. However,

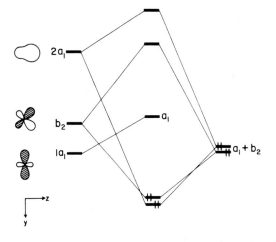

Fig. 17. Partial MO Scheme of Cp_2TiX_2: On the left the MO's of the fragment Cp_2Ti, on the right the orbitals of two σ-donors [reproduced with permission from J. W. Lauher and R. Hoffmann, J. Am. Chem. Soc. **98**, 1729 (1976)]

fairly large variations in $\varnothing$ do not cause great changes in total energy (at $\varnothing = 140°$, only 0.4 eV higher). The experimental value of $\varnothing$ in *96* is 148° [201]. The authors conclude that the formal 16-electron molecule is really an 18-electron molecule due to effective π-donation by the two oxygen ligands [201]. This interaction is believed to involve the $1a_1$ orbital of the Cp_2Ti-fragment (Fig. 17) as shown in *98*.

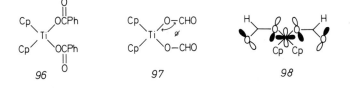

In an interesting study involving X-ray photoelectron spectroscopy (ESCA) as well as electrochemical measurements on a number of Cp_2TiX_2 compounds, Gassman has renewed the debate concerning bonding in Cp_2TiCl_2 [199]. A large electronic effect is observed upon replacing the Cp-groups by pentamethyl-cyclopentadienyl analogs (*99*), as reflected by the binding energies of the inner-shell electrons of the metal and by the oxidation potentials. The methyl groups exert an effect which is equivalent to a one-electron reduction of titanium! The pronounced electronic effect of methyl substitution is also revealed by Ti-NMR spectroscopy [180], as described earlier. Also, within a series, the oxidation potential $e_{1/2}$ does not vary in going from the dichloride to the dibromide. Since the HOMO of Cp_2TiCl_2 cannot be associated with the metal (as previously deduced), oxidation must involve the Cp ligands [199]. In contrast to Dahl's suggestion that the HOMO involves the chlorine lone electron pairs [197], Gassman concludes that the HOMO is Cp based [180]. This seems to be the currently accepted view [202].

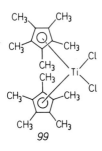

99

In a detailed self-consistent field-X_α-scattered wave (SCF—X_α—SW) molecular orbital study of Cp_2TiX_2 (X = F, Cl, Br, I, CH_3), Tyler has also reconsidered the bonding in such compounds [202]. The lowest energy electronic transition of Cp_2TiCl_2 was predicted to be the Cp → Ti charge-transfer transition. According to the calculation, the Cl → Ti charge-transfer transition occurs at higher energy. The calculated ionization energies compare fairly well with the experimental data. The eigenvalues and wave function contour plots of Cp_2TiCl_2 were also computed in this study [202].

The most pronounced substituent effect in bis(cyclopentadienyl)titanium dichlorides to date has been observed by Boche and co-workers [203]. They synthesized the novel *bis*-amino derivative *100*. Whereas Cp_2TiCl_2 is red-orange, *100* turned out to be green-black. The crystal structure shows an unusually short Cp—N bond distance (1.347 Å), which indicates that the nitrogen lone-pair is "strongly engaged in stabilizing the 16-electron Ti complex" [203]. Indeed, *100* resists methylation with CH_3I or Meerwein reagent. Cyclic voltammetry also indicates a large electronic effect due to the electron-donating nitrogen: The reduction potentials E° amount to —1.11 V (SCE) and —1.84 V (SCE) [203], compared to —0,64 and —2.06 V for the unsubstituted Cp_2TiCl_2 [204a].

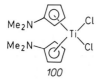

100

The PE spectra of $CpTiX_3$ (X=Cl, Br) and some methycyclopentadienyl and pentamethylcyclopentadienyl analogs have been studied recently [205]. All of the spectra display a band at lowest energy which shifts strongly to lower vertical ionization values upon methyl substitution: *43* (9.79 eV), *101* (9.61 eV) and *102* (8.87 eV). Consequently, these IP's were assigned to the Cp-π orbital. A detailed analysis shows that these bands also have a small $TiCl_3$-character [205]. The second band in all the spectra appears to be due to ionization from the a_2-in-plane-Cl lone pair orbital; the IP's decrease in the series *43* (10.77 eV), *101* (10.66 eV) and *102* (10.39 eV). Higher energy bands were also analyzed. Finally, mixing between Cp and d-type orbitals at titanium was found to be strong and essentially independent of the nature of the halogen (Cl or Br) [205]. The MO-diagram of *43* as derived from the PE data and MO calculations is shown in Fig. 18.

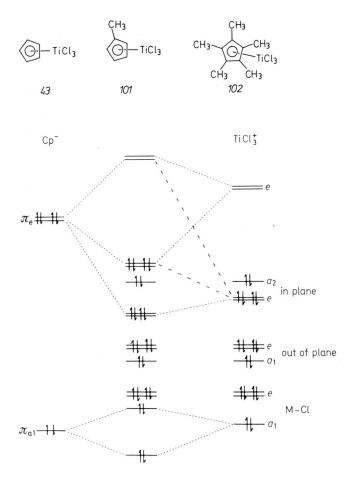

Fig. 18. MO Diagram of CpTiCl$_3$ [reproduced with permission from A. Terpstra, J. N. Lowen, A. Oskam and J. H. Teuben, J. Organomet. Chem. *260*, 207 (1984)]

The investigations of CpTiX$_3$ and Cp$_2$TiCl$_2$ and their derivatives clearly point to large substituent effects in the Cp-ligand. In considering the adjustment of carbanion-selectivity via titanation, the choice of substituents of a Cp ligand thus offers another way to influence the electronic property of titanium (Chapter 1).

Besides the previously discussed theoretical papers, several other molecular orbital studies of various aspects of titanium chemistry have appeared [206].

The pharmacological properties of organotitanium compounds have not been studied systematically. However, Köpf and Maier-Köpf have shown that Cp_2TiCl_2 and similar compounds constitute an interesting class of anti-tumor agents [207].

2.6 Conclusions

Some of the above physical, chemical and computational data is useful in applying the principle of adjusting carbanion-selectivity via titanation. The following chapters deal primarily with synthetic aspects. In a few cases additional structural and spectroscopic data are available, but more work is necessary. Indeed, the precise structure, aggregation state and electronic nature of many new and synthetically useful organotitanium reagents remain to be explored. Until then, simple analogy based on the information provided in this chapter serves as the best guide.

References

1. Reviews of Ti(II), Ti(III) and Ti(IV) Compounds: a) Gmelin Handbuch, "Titan-Organische Verbindungen", Part 1 (1977), Part 2 (1980), Part 3 (1984), Part 4 (1984), Springer-Verlag, Berlin; b) Segnitz, A.: in Houben-Weyl-Müller, "Methoden der Organischen Chemie", Vol. 13/7, p. 261, Thieme-Verlag, Stuttgart 1975; c) Wailes, P. C., Coutts, R. S. P., Weigold, H.: "Organometallic Chemistry of Titanium, Zirconium and Hafnium", Academic Press, N.Y. 1974; d) Labinger, J. A.: J. Organomet. Chem. *227*, 341 (1982); e) Bottrill, M., Gavens, P. D., Kelland, J. W., McMeeking, J.: in "Comprehensive Organometallic Chemistry", (Wilkinson, G., Stone, F. G. A., Abel, E. W., editors), Chapter 22, Pergamon Press, Oxford (1982).
2. Cahours, M. A.: Ann. Chim. (Paris) *62*, (3) 280 (1861).
3. Schumann, A.: Ber. dtsch. chem. Ges. *21*, 1080 (1888).
4. Paterno, E., Peratoner, A.: Ber. dtsch. chem. Ges. *22*, 467 (1889).
5. a) Levy, M. L.: Ann. Chim. (Phys.) *25*, (6) 433 (1892); see also b) Köhler, H., Ber. dtsch. chem. Ges. *13*, 1626 (1880).
6. Challenger, F., Pritchard, F.: J. Chem. Soc. *125*, I, 864 (1924).
7. a) Plets, V. M.: J. Gen. Chem. (Moscow) *8*, 1298 (1938); b) Browne, O. H., Reid, E. E.: J. Am. Chem. Soc. *49*, 830 (1927); c) Razuvaev, G. A., Bodanov, I. F.: Zh. Obshch. Khim. *8*, 1298 (1938).
8. Gilman, H., Jones, R. G.: J. Org. Chem. *10*, 505 (1945).
9. a) Herman, D. F., Nelson, W. K.: J. Am. Chem. Soc. *74*, 2693 (1952); b) Herman, D. F., Nelson, W. K.: J. Am. Chem. Soc. *75*, 3877 (1953).
10. Holloway, H.: Chem. Ind. *1962*, 214.
11. a) Dijkgraaf, C., Rousseau, J. P. G.: Spectrochim. Acta *24A*, 1213 (1968); b) Reetz, M. T., Steinbach, R., Kesseler, K.: Angew. Chem. *94*, 872 (1982); Angew. Chem., Int. Ed. Engl. *21*, 864 (1982); Angew. Chem. Supplement *1982*, 1899.
12. a) Herman, D. F.: U.S. Patent 29 63 471 (1960); b) Razuvaev, G. A., Bobinova, L. A., Etlis, V. S.: Dokl. Akad. Nauk. SSSR *127*, 581 (1959); c) North, A. M.: Proc. Roy. Soc. [London] *254A*, 408 (1960).

13. a) Herman, D. F.: U.S. Patent 28 86 579 (1959); b) Herman, D. F.: Adv. Chem. Ser. Nr. 23, p. 265 (1959).
14. Herman, D. F., Nelson, W. K.: J. Am. Chem. Soc. *75*, 3882 (1953).
15. Reviews: a) Sinn, H., Kaminsky, W.: Adv. Organomet. Chem. *18*, 99 (1980); Pino, P., Mülhaupt, R.: Angew. Chem. *92*, 869 (1980); Angew. Chem., Int. Ed. Engl. *19*, 857 (1980); c) see also ref. [1*e*].
16. a) Wilkinson, G., Rosenblum, H., Whiting, M. C., Woodward, R. B.: J. Am. Chem. Soc. *74*, 2125 (1952); b) Fischer, E. O., Pfab, W.: Z. Naturforsch. *7b*, 377 (1952).
17. a) Wilkinson, G., Pauson, P. L., Birmingham, J. M., Cotton, F. A.: J. Am. Chem. Soc. *75*, 1011 (1953); b) review: Pez, G. P., Armor, J. N.: Adv. Organomet. Chem. *19*, 1 (1981).
18. a) Beermann, C.: Angew. Chem. *71*, 195 (1959); b) Bestian, H., Clauss, K., Jensen, H., Prinz, E.: Angew. Chem. *74*, 955 (1962); Angew. Chem., Int. Ed. Engl. *2*, 32 (1963).
19. Clauss, K.: Liebigs Ann. Chem. *711*, 19 (1968).
20. Rausch, M. D., Gordon, H. B.: J. Organomet. Chem. *74*, 85 (1974).
21. Westermann, J.: Dissertation, Univ. Marburg 1982.
22. Reetz, M. T., Steinbach, R., Westermann, J., Peter, R.: Angew. Chem. *92*, 1011 (1980); Angew. Chem., Int. Ed. Engl. *19*, 1044 (1980).
23. de Vries, H.: Rec. Trav. Chim. Pays-Bas *80*, 866 (1961).
24. a) Thiele, K. H., Zdunneck, P., Baumgart, D.: Z. Anorg. Allg. Chem. *378*, 62 (1970); b) Fowles, G. W. A., Rice, D. A., Wilkins, J. D.: J. Chem. Soc. A, *1971*, 1920); c) Gray, A. P.: Can. J. Chem. *41*, 1511 (1963).
25. Bawn, C. E. H., Gladstone, J.: Proc. Chem. Soc. *1959*, 227.
26. Dijkgraaf, C., Rousseau, J. P. G.: Spectrochim. Acta *25*, 1455 (1969).
27. Kyung, S. H.: Dissertation, Univ. Marburg 1985.
28. Reetz, M. T., Westermann, J., Kyung, S. H.: Chem. Ber., *118*, 1050 (1985).
29. Reetz, M. T., Westermann, J., Steinbach, R.: Angew. Chem. *92*, 931 (1980); Angew. Chem., Int. Ed. Engl. *19*, 900 (1980).
30. a) McCowan, J. D.: Can. J. Chem. *51*, 1083 (1973); b) van Looy, H. M., Rodriguez, L. A. M., Gabant, J. A.: J. Polym. Sci. Polymer Chem. Ed. *4*, 1927 (1966).
31. Edgecombe, F. H. C.: Can. J. Chem. *41*, 1265 (1963).
32. McCowan, J. D., Hanlan, J. F.: Can. J. Chem. *50*, 755 (1972).
33. Lit. [1a], Part 1, pp. 29ff.
34. Thiele, K. H.: Pure Appl. Chem. *30*, 575 (1972).
35. Dong, D., Stevens, S. C. V., McCowan, J. D., Baird, M. C.: Inorg. Chim. Acta *29*, L225 (1978).
36. Clark, R. J. H., McAlees, A. J.: J. Chem. Soc. A., *1970*, 2026.
37. Kyung, S. H.: Diplomarbeit, Univ. Marburg 1983.
38. Reetz, M. T., Westermann, J.: Synth. Commun. *11*, 647 (1981).
39. Dawoodi, Z., Green, M. L. H., Mtetwa, V. S. B., Prout, K.: J. Chem. Soc., Chem. Commun. *1982*, 1410.
40. Brookhart, M., Green, M. L. H.: J. Organomet. Chem. *250*, 395 (1983).
41. Clark, R. J. H., Coles, M. A.: J. Chem. Soc., Dalton Trans. *1972*, 2454.
42. a) Bawn, C. E. H., Symcox, R.: J. Polym. Sci. *34*, 139 (1959); b) Takeda, M., Iimura, K., Nozawa, Y., Hisatome, M., Koide, N.: J. Polym. Sci. Polymer Symposium Nr. *23*, 741 (1968).
43. Lappert, M. F., Patil, D. S., Pedley, J. B.: J. Chem. Soc., Chem. Commun. *1975*, 830.
44. Wilkinson, G.: Pure Appl. Chem. *30*, 627 (1972).

45. Reviews of homoleptic metal σ-hydrocarbyls: a) Davidson, P. J., Lappert, M. F., Pearce, R.: Chem. Rev. 76, 219 (1976); b) Schrock, R. R., Parshall, G. W.: Chem. Rev. 76, 243 (1976).
46. Cooper, G. D., Finkbeiner, H. L.: J. Org. Chem. 27, 1493 (1962).
47. a) Fell, B., Asinger, F., Sulzbach, R. A.: Chem. Ber. 103, 3830 (1970); b) Sato, F.: J. Organomet. Chem. 285, 53 (1985).
48. Wenderoth, B.: Diplomarbeit, Univ. Marburg 1980.
49. a) Reetz, M. T., Steinbach, R., Wenderoth, B., Westermann, J.: Chem. Ind. 1981, 541; b) Weidmann, B., Widler, L., Olivero, A. G., Maycock, C. D., Seebach, D.: Helv. Chim. Acta 64, 357 (1981); c) Reetz, M. T., Steinbach, R., Westermann, J., Urz, R., Wenderoth, B., Peter, R.: Angew. Chem. 94, 133 (1982); Angew. Chem., Int. Ed. Engl. 21, 135 (1982); Angew. Chem. Supplement 1982, 257.
50. Reviews of recent applications of organotitanium chemistry in organic synthesis: a) Reetz, M. T.: Top. Curr. Chem. 106, 1, (1982); b) Weidmann, B., Seebach, D.: Angew. Chem. 95, 12 (1983); Angew. Chem., Int. Ed. Engl. 22, 31 (1983); c) Reetz, M. T.: Pure Appl. Chem. 57, 1781 (1985).
51. Thiele, K. H., Jacob, K.: Z. Anorg. Allg. Chem. 356, 195 (1968).
52. Beermann, C., Bestian, H.: Angew. Chem. 71, 618 (1959).
53. Dawoodi, Z., Green, M. L. H., Mtetwa, V. S. B., Prout, K.: J. Chem. Soc., Chem. Commun. 1982, 802.
54. Langguth, E., Van Thu, N., Shakoor, A., Jacob, K., Thiele, K. H.: Z. Anorg. Allg. Chem. 505, 127 (1983).
55. Weidmann, B., Maycock, C. D., Seebach, D.: Helv. Chim. Acta 64, 1552 (1981).
56. a) Beermann, C.: German Patent D.P. 11 00 022 (1961); b) Ivanova, V. P., Bresler, L. S., Dolgoplosk, B. A.: Izv. Akad. Nauk. SSSR, Ser. Khim. 1979, 2742.
57. a) Hewitt, B. J., Holliday, A. K., Puddephatt, R. J.: J. Chem. Soc., Dalton Trans. 1973, 801; b) Cardin, D. J., Norton, R. J.: J. Chem. Soc., Chem. Commun. 1979, 513.
58. Reetz, M. T., Westermann, J., Steinbach, R., Wenderoth, B., Peter, R., Ostarek, R., Maus, S.: Chem. Ber. 118, 1421 (1985).
59. a) Beilin, S. I., Bondarenko, G. N., Vdovin, V. M., Dolgoplosk, B. A., Markevich, I. N., Nametkin, N. S., Poletaev, V. A., Svergun, V. I., Sergeeva, M. B.: Dokl. Akad. Nauk. SSSR 218, 1347 (1974); b) Sonnek, G., Baumgarten, K. G., Reinheckel, H., Schroeder, S., Thiele, K. H.: Z. Anorg. Allg. Chem. 426, 232 (1976); c) Middleton, A. R., Wilkinson, G.: J. Chem. Soc. Dalton Trans. 1980, 1888.
60. Razuvaev, G. A., Latyaeva, V. N., Kilyakova, G. A.: Dokl. Chem. Proc. Acad. Sci. USSR 203, 220 (1972).
61. Seidel, W., Riesenberg, E.: Z. Chem. 20, 450 (1980).
62. Wenderoth, B.: Dissertation, Univ. Marburg 1983.
63. a) Bürger, H., Neese, H. J.: Chimia 24, 209 (1970); b) Bürger, H., Neese, H. J.: J. Organomet. Chem. 20, 129 (1969); c) Bürger, H., Neese, H. J.: J. Organomet. Chem. 21, 381 (1970); d) Bürger, H., Neese, H. J.: J. Organomet. Chem. 36, 101 (1976).
64. Fowles, G. W. A., Hoodless, R. A.: J. Chem. Soc. 1963, 33.
65. Kühlein, K., Clauss, K.: Makromol. Chem. 155, 145 (1972).
66. Zucchini, U., Albizzati, E., Giannini, U.: J. Organomet. Chem. 26, 357 (1971).
67. Reetz, M. T., et al., unpublished.
68. Gumboldt, A., Jastrow, H.: German Patent, DBP 12 03 775 (1965).
69. Sugahara, H., Shuto, Y.: J. Organomet. Chem. 24, 709 (1970).

70. Meyer, E. M., Jacot-Guillarmod, A.: Helv. Chim. Acta *66*, 898 (1983), and previous papers cited therein.

71. Hüllmann, M.: projected Dissertation, Univ. Marburg.

72. Hanlan, J. F., McCowan, J. D.: Can. J. Chem. *50*, 747 (1972).

73. Clark, R. J. H., Coles, M. A.: J. Chem. Soc. Dalton Trans. *1974*, 1462.

74. a) Guyot, A., Peyroche, J., Laputte, R.: Kinet. Mech. Polyreactions, Int. Symp. Macromol. Chem. Budapest, Preprints Vol. 2, p. 199 (1971); b) see lit. [1a], Part 1, p. 72.

75. Bürger, H., Kluess, C.: J. Organomet. Chem. *108*, 69 (1976).

76. D'Alelio, G. F.: US Patent, USP 30 60 161 (1962).

77. Berthold, H. J., Groh, G.: Angew. Chem. *75*, 576 (1963).

78. Bresler, L. S., Poddubnyi, I. Y., Smirnova, T. K., Khachaturov, A. S., Tsereteli, I. Y.: Dokl. Akad. Nauk. SSSR *210*, 847 (1973).

79. a) Jacot-Guillarmod, A., Tabacchi, R., Causse, J.: Chimia *23*, 188 (1969); b) Tabacchi, R., Jacot-Guillarmod, A.: Chimia *25*, 326 (1971).

80. a) Clauss, K., Beermann, C.: Angew. Chem. *71*, 627 (1959); b) Berthold, H. J., Groh, G.: Z. Anorg. allg. Chem. *319*, 230 (1963); c) Thiele, K. H., Müller, J.: J. Prakt. Chem. *38*, 147 (1968); d) Tabacchi, R., Jacot-Guillarmod, A.: Helv. Chim. Acta *53*, 1977 (1970); e) Rau, H., Müller, J.: Z. Anorg. Allg. Chem. *415*, 225 (1975).

81. For synthetic applications of $(CH_3)_4Ti$, $TiCl_4$ was treated with CH_3Li [49c]; CH_3MgCl can also be used [67].

82. See p. 235 of ref. [45a].

83. a) Dubsky, G. J., Boustany, K. S., Jacot-Guillarmod, A.: Chimia *24*, 17 (1970); b) Boustany, K. S., Bernauer, K., Jacot-Guillarmod, A.: Helv. Chim. Acta *50*, 1305 (1967); c) Wilke, G., Bogdanovic, B., Hardt, P., Heimbach, P., Keim, W., Kröner, M., Oberkirch, W., Tanaka, K., Steinrücke, E., Walter, D., Zimmermann, H.: Angew. Chem. *78*, 157 (1966); Angew. Chem., Int. Ed. Engl. *5*, 151 (1966).

84. a) Boustany, K. S., Bernauer, K., Jacot-Guillarmod, A.: Helv. Chim. Acta *50*, 1080 (1967); b) Giannini, U., Zucchini, U.: J. Chem. Soc., Chem. Commun. *1968*, 940; c) Brüser, W., Thiele, K. H., Zdunneck, P., Brune, F.: J. Organomet. Chem. *32*, 335 (1971).

85. a) Bassi, I. W., Allegra, G., Scordamaglia, R., Chioccola, G.: J. Am. Chem. Soc. *93*, 3787 (1971); b) Davies, G. R., Jarvis, J. A. J., Kilbourn, B. T.: J. Chem. Soc., Chem. Commun. *1971*, 1511.

86. Tabacchi, R., Boustany, K. S., Jacot-Guillarmod, A.: Helv. Chim. Acta *53*, 1971 (1970).

87. Collier, M. R., Lappert, M. F., Pearce, R.: J. Chem. Soc., Dalton Trans. *1973*, 445.

88. a) Davidson, P. J., Lappert, M. F., Pearce, R.: J. Organomet. Chem. *57*, 269 (1973); b) Mowat, W., Wilkinson, G.: J. Chem. Soc. Dalton Trans. *1973*, 1120.

89. Bower, B. K., Tennent, H. G.: J. Am. Chem. Soc. *94*, 2512 (1972).

90. Roberts, R. M. G.: J. Organomet. Chem. *63*, 159 (1973).

91. Bochmann, M., Wilkinson, G., Young, G. B.: J. Chem. Soc., Dalton Trans. *1980*, 1879.

92. a) Müller, J., Rau, H., Zdunnek, P., Thiele, K. H.: Z. Anorg. Allg. Chem. *401*, 113 (1973); b) see also ref. [1a], Part 1, p. 121.

93. Röder, A., Scholz, J., Thiele, K. H.: Z. Anorg. Allg. Chem. *505*, 121 (1983).

94. a) Wieghardt, K., Tolksdorf, I., Weiss, J., Swiridoff, W.: Z. Anorg. Allg. Chem. *490*, 182 (1982); b) for other examples of titanium compounds having

coordination numbers higher than six, see: Clark, R. J. H.: in "Comprehensive Inorganic Chemistry", Vol. III, p. 355ff., Pergamon Press 1973.

95. a) Semin, G. K., Nogina, O. V., Dubovitskii, V. A., Babushkina, T. A., Bryukhova, E. V., Nesmeyanov, A. N.: Dokl. Chem. (Engl. Trans.) *194*, 635 (1970); b) Cardoso, A. M., Clark, R. J. H., Moorhouse, S.: J. Chem. Soc. Dalton Trans. *1980*, 1156; c) see also ref. [1e], p. 332.

96. a) Wilkinson, G., Birmingham, J. M.: J. Am. Chem. Soc. *76*, 4281 (1954); b) Summers, L., Uloth, R. H., Holmes, A.: J. Am. Chem. Soc. *77*, 3604 (1955); c) see also ref. [1e], p. 361.

97. Calderon, J. L., Cotton, F. A., De Boer, B. G., Takats, J.: J. Am. Chem. Soc. *93*, 3592 (1971).

98. Clark, R. J. H., Stockwell, J. A., Wilkins, J. D.: J. Chem. Soc. Dalton Trans. *1976*, 120.

99. a) Giannini, U., Cesca, S.: Tetrahedron Lett. *1*, 19 (1960); b) Clauss, K., Bestian, H.: Liebigs Ann. Chem. *654*, 8 (1962); c) Green M. L. H., Lucas, C. R.: J. Organomet. Chem. *73*, 259 (1974).

100. a) Piper, T. S., Wilkinson, G.: J. Inorg. Nucl. Chem. *3*, 104 (1956); b) Bercaw, J. E., Brintzinger, H. H.: J. Am. Chem. Soc. *91*, 7301 (1969); c) Alt, H. G., Di Sango, F. P., Rausch, M. D., Uden, P. C.: J. Organomet. Chem. *107*, 257 (1976); d) Sinn, H., Patat, F.: Angew. Chem. *75*, 805 (1963); e) McDermott, J. X., Wilson, M. E., Whitesides, G. M.: J. Am. Chem. Soc. *98*, 6529 (1976); f) Boekel, C. P., Teuben, J. H., De Liefde Meijer, H. J.: J. Organomet. Chem. *102*, 317 (1975); g) Wozniak, B., Ruddick, J. D., Wilkinson, G.: J. Chem. Soc. A, *1971*, 3116; h) Beachell, H. C., Butter, S. A.: Inorg. Chem. *4*, 1133 (1965); see also ref. [1e], p. 396ff. for a more extensive discussion of these compounds.

101. Waters, J. A., Mortimer, G. A.: J. Organomet. Chem. *22*, 417 (1970).

102. a) Khan, O., Dormond, A., Letourneux, J. P.: J. Organomet. Chem. *132*, 149 (1977); b) Lappert, M. F., Martin, T. R., Atwood, J. L., Hunter, W. E.: J. Chem. Soc., Chem. Commun. *1980*, 476; see also ref. [1e], p. 408ff.

103. Facchinetti, G., Floriani, C.: J. Organomet. Chem. *71*, C5 (1974).

104. a) Murray, J. G.: J. Am. Chem. Soc. *83*, 1287 (1961); b) Siegert, F. W., De Liefde Meijer, H. J.: J. Organomet. Chem. *20*, 141 (1969); c) Masai, H., Sonogashira, K., Hagihara, N.: Bull. Chem. Soc. Jap. *41*, 750 (1968).

105. Floriani, C., Fachinetti, G.: J. Chem. Soc., Chem. Commun. *1972*, 790.

106. a) Rausch, M. D.: Pure Appl. Chem. *30*, 523 (1972); b) Mattia, J., Humphrey, M. B., Rogers, R. D., Atwood, J. L., Rausch, M. D.: Inorg. Chem. *17*, 3257 (1978).

107. McDermott, J. X., Whitesides, G. M.: J. Am. Chem. Soc. *96*, 947 (1974).

108. Grubbs, R. H., Miyashita, A.: J. Am. Chem. Soc. *100*, 1300 (1978).

109. Woodward, R. B., Hoffmann, R.: Angew. Chem. *81*, 797 (1969); Angew. Chem., Int. Ed. Engl. *8*, 781 (1969).

110. a) Sakurai, H., Umino, H.: J. Organomet. Chem. *142*, C49 (1977); b) Kira, M., Bock, H., Umino, H., Sakurai, H.: J. Organomet. Chem. *173*, 39 (1979).

111. Ho, S. C. H., Straus, D. A., Grubbs, R. H.: J. Am. Chem. Soc. *106*, 1533 (1984); and lit. cited therein (see also Chapter 8).

112. Tebbe, F. N., Parshall, G. W., Reddy, G. S.: J. Am. Chem. Soc. *100*, 3611 (1978).

113. Pine, S. H., Zahler, R., Evans, D. A., Grubbs, R. H.: J. Am. Chem. Soc. *102*, 3272 (1980).

114. Temple, J. S., Riediker, M., Schwartz, J.: Am. Chem. Soc. *104*, 1310 (1982); and lit. cited therein.

115. Teuben, J. H.: J. Organomet. Chem. *57*, 159 (1973).
116. a) Brintzinger, H. H.: J. Am. Chem. Soc. *89*, 6871 (1967); b) Bercaw, J. E., Brintzinger, H. H.: J. Am. Chem. Soc. *91*, 7301 (1969).
117. a) Bercaw, J. E.: J. Am. Chem. Soc. *96*, 5087 (1974); b) Breil, A., Wilke, G.: Angew. Chem. *20*, 942 (1966); Angew. Chem., Int. Ed. Engl. *5*, 898 (1966); c) Schwartz, J., Sadler, J. E.: J. Chem. Soc., Chem. Commun. *1973*, 172.
118. a) Anthony, M. T., Green, M. L. H., Young, D.: J. Chem. Soc. Dalton Trans. *1975*, 1419; b) Wawker, P. N., Kuendig, P. E., Timms, P. L.: J. Chem. Soc., Chem. Commun. *1978*, 730.
119. Tel'noi, V. I., Rabinovich, I. B., Tikhonov, V. D., Latyeava, V. N., Yyshinskaya, L. I., Razuvaev, G. A.: Dokl. Akad. Nauk SSSR *174*, 1374 (1967).
120. a) Calhorda, M. J., Dias, A. R., Salema, M. S., Simoes, J. A. M.: J. Organomet. Chem. *255*, 81 (1983); b) Kochi, J. K.: Pure Appl. Chem. *52*, 571 (1980).
121. For a more detailed discussion of decomposition pathways, see ref. [1e], p. 459 ff.
122. Khachaturov, A. S., Bresler, L. S., Poddubnyi, I. Y.: J. Organomet. Chem. *64*, 335 (1974); and lit. cited therein.
123. Morino, Y., Uehara, H.: J. Chem. Phys. *45*, 4543 (1966).
124. Brand, P., Sackmann, H.: Z. Anorg. Allg. Chem. *321*, 262 (1963).
125. Crystal packing effects appear not to be involved [85].
126. Davies, G. R., Jarvis, J. A. J., Kilbourn, B. T., Pioli, A. J. P.: J. Chem. Soc., Chem. Commun. *1971*, 677.
127. Davies, G. R., Jarvis, J. A. J., Kilbourn, B. T.: J. Chem. Soc., Chem. Commun. *1971*, 1511.
128. Brauer, D. J., Bürger, H. H., Wiegel, K.: J. Organomet. Chem. *150*, 215, (1978).
129. Guryanova, E. N., Goldshtein, I. P., Romm, I. P.: "Donor-Acceptor Bond", Wiley, N.Y., 1975.
130. Bose, A. K., Srinivasan, P. R., Trainor, G.: J. Am. Chem. Soc. *96*, 3670 (1974).
131. Kletenik, Y. B., Osipov, O. A.: Zh. Obshch. Khim. *31*, 710 (1961).
132. Brun, L.: Acta Cryst. *20*, 739 (1966).
133. a) Bränden, C. I., Lindqvist, I.: Acta Chem. Scand. *14*, 726 (1960); b) Bassi, I. W., Calcaterra, M., Intrito, R.: J. Organomet. Chem. *127*, 305 (1977).
134. Poll, T., Metter, J. O., Helmchen, G.: Angew. Chem. *97*, 116 (1985); Angew. Chem., Int. Ed. Engl. *24*, 112 (1985).
135. a) Ganis, P., Allegra, G.: Atti Acad. Nazl. Lincei Rend. Classe Sci. Fis. Mat. Nat. *33*, 303 (1962); b) Engelhardt, L. M., Papasergio, R. I., Raston, C. L., White, A. H.: Organometallics *3*, 18 (1984).
136. Clearfield, A., Warner, D. K., Saldarriaga-Molina, C. H., Ropal, R., Bernal, I.: Can. J. Chem. *53*, 1622 (1975).
137. McKenzie, T. C., Sanner, R. D., Bercaw, J. E.: J. Organomet. Chem. *102*, 457 (1975).
138. Huffmann, J. C., Moloy, K. G., Marsella, J. A., Caulton, K. G.: J. Am. Chem. Soc. *102*, 3009 (1980).
139. Besançon, J., Top, S., Tirouflet, J., Dusausoy, Y., Lecomte, C., Protas, J.: J. Organomet. Chem. *127*, 153 (1977).
140. Gray, A. P., Callear, A. B., Edgecombe, F. H. C.: Can. J. Chem. *41*, 1502 (1963).
141. Coutts, R. S. P., Wailes, P. C.: Aust. J. Chem. *24*, 1075 (1971).
142. Besançon, J., Top, S., Tirouflet, J., Dusausoy, J., Lecomte, C., Protas, J.: J. Chem. Soc., Chem. Commun. *1976*, 325.

143. a) Siegert, F. W., De Liefde Meijer, H. J.: Rec. Trav. Chim. Pays-Bas *89*, 764 (1970); b) Atwood, J. L., Hunter, W. E., Alt, H., Rausch, M. D.: J. Am. Chem. Soc. *98*, 2454 (1976).
144. Peter, R.: Dissertation, Univ. Marburg 1983.
145. Bradley, D. C., Mehrota, R. C., Wardlaw, W.: J. Chem. Soc. *1952*, 5020.
146. Bradley, D. C., Mehrota, R. C., Gaur, D. P.: "Metal Alkoxides", Academic Press, London 1978.
147. Wright, D. A., Williams, D. A.: Acta Crystallogr. *B24*, 1107 (1968).
148. Masthoff, R., Köhler, H., Böhland, H., Schmeil, F.: Z. Chem. *5*, 122 (1965).
149. Ibers, J. A.: Nature *197*, 686 (1963).
150. a) Caughlan, C. N., Smith, H. S., Katz, W., Hodgson, W., Crowe, R. W.: J. Am. Chem. Soc. *73*, 5652 (1951); b) Martin, R. L., Winter, G.: J. Chem. Soc. *1961*, 2947.
151. Russo, W. R., Nelson, W. H.: J. Am. Chem. Soc. *92*, 1521 (1970).
152. a) Bradley, D. C., Holloway, C. E.: J. Chem. Soc. A *1968*, 1316; Holloway, C. E.: J. Chem. Soc. Dalton Trans. *1976*, 1050.
153. Barraclough, C. G., Martin, R. L., Winter, G.: J. Chem. Soc. *1964*, 758.
154. Cullinane, N. M., Chard, S. J., Price, G. F., Millward, B. B., Langlois, G.: J. Appl. Chem. *1*, 400 (1951).
155. a) Watenpaugh, K., Caughlan, C. N.: Inorg. Chem. *5*, 1782 (1966); b) Frazer, M. J., Goffer, Z.: J. Inorg. Nucl. Chem. *28*, 2410 (1966).
156. a) Crystal structure: Weller, H. O., Müller, U.: Chem. Ber. *109*, 3039 (1976); b) synthesis: Dehnicke, K.: J. Inorg. Nucl. Chem. *27*, 809 (1965).
157. Our own cryoscopic measurements substantiate the results reported in ref. [65]. In benzene at a concentration of 0.11 M at ~ 5 °C we found a molecular weight of 346.
158. Maus, S.: projected Dissertation, Univ. Marburg 1986.
159. Flamini, A., Cole-Hamilton, D. J., Wilkinson, G.: J. Chem. Soc. Dalton Trans. *1978*, 454.
160. Yoshino, A., Shuto, Y., Iitaka, Y.: Acta Crystallogr. *B26*, 744 (1970).
161. Stoeckli-Evans, H.: Helv. Chim. Acta *58*, 373 (1975).
162. Williams, I. D., Pedersen, S. F., Sharpless, K. B., Lippard, S. J.: J. Am. Chem. Soc. *106*, 6430 (1984).
163. Scharf, W., Neugebauer, D., Schubert, U., Schmidbaur, H.: Angew. Chem. *90*, 628 (1978); Angew. Chem., Int. Ed. Engl. *17*, 601 (1978).
164. Begley, J. W., Pennella, F.: J. Catalysis *8*, 203 (1967).
165. Roshchupkina, O. S., Vinogradova, S. M., Borodko, Y. G.: Russ. J. Phys. Chem. *46*, 1553 (1972).
166. Armstrong, D. R., Perkins, P. G., Steward, J. J. P.: Rev. Roumaine Chim. *20*, 177 (1975).
167. Streitwieser, A., Williams, J. E., Alexandratos, S., McKelvey, J. M.: J. Am. Chem. Soc. *98*, 4778 (1976).
168. a) Bachrach, S. M., Streitwieser, A.: J. Am. Chem. Soc. *106*, 2283 (1984); b) Schleyer, P. v. R., Kos, A. J., Kauffmann, E.: J. Am. Chem. Soc. *105*, 7617 (1983); c) Streitwieser, A.: Acc. Chem. Res. *17*, 353 (1984).
169. Basso-Bert, M., Cassoux, P., Crasnier, F., Gervais, D., Labarre, J. F., De Loth, P.: J. Organomet. Chem. *136*, 201 (1977).
170. PE-Studies of $TiCl_4$: a) Cox, P. A., Evans, S., Hamnet, A., Orchard, A. F.: Chem. Phys. Lett. *7*, 414 (1970); b) Bancroft, G. M., Pellach, E., Tse, J. S.: Inorg. Chem. *21*, 2950 (1982), and lit. cited therein concerning earlier PE studies of $TiCl_4$.
171. Similar observations have been made in going from CD_2Cl_2 to C_6D_6 [158].

172. Dijkgraaf, C.: J. Phys. Chem. *69*, 660 (1965).
173. a) Pritchard, H. O., Skinner, H. A.: Chem. Rev. *55*, 745 (1955); b) Drago, R. S., Matwiyoff, N. A.: J. Organomet. Chem. *3*, 62 (1965); c) Huheey, J. E.: "Inorganic Chemistry", 3rd ed., p. 845ff., Harper and Row, N.Y. 1983; d) Mullay, J.: J. Am. Chem. Soc. *106*, 5842 (1984).
174. Blandy, C., Guerreiro, R.: Compt. Rend. C *278*, 1323 (1974).
175. Beachell, H. C., Butter, S. A.: Inorg. Chem. *4*, 1133 (1965).
176. Erskine, G. J., Hurst, G. J. B., Weinberg, E. L., Hunter, B. K., McCowan, J. D.: J. Organomet. Chem. *267*, 265 (1984).
177. Kidd, R. G., Goodfellow, R. J.: in "NMR and the Periodic Table", (Harris, R. K., Mann, B. E., editors), p. 201ff., Academic Press, N.Y. 1978.
178. Kidd, R. G., Matthews, R. W., Spinney, H. G.: J. Am. Chem. Soc. *94*, 6686 (1972).
179. Hav, N., Sayer, B. G., Denes, G., Bickley, D. G., Detellier, C., McGlinchey, M. J.: J. Magn. Reson. *50*, 50 (1982).
180. a) Gassman, P. G., Cambell, W. H., Macomber, D. W.: Organometallics *3*, 385 (1984); b) Dormond, A., Fauconet, M., Leblanc, J. C., Moise, C.: Polyhedron *3*, 897 (1984)
181. Clark, R. J. H., Willis, C. J.: Inorg. Chem. *10*, 1118 (1971).
182. a) Karapinka, G. L., Smith, J. J., Carrick, W. L.: J. Polym. Sci. *50*, 143 (1961); b) Groenewege, M. P.: Z. Physik. Chem. *18*, 147 (1958); c) for a detailed discussion of the IR bands of CH_3TiCl_3, see ref. [1a], Part 1, p. 18ff.
183. la Lau, C.: Rec. Trav. Chim. Pays-Bas *84*, 429 (1965).
184. a) Bencivenni, L., Gigli, R., Pelino, M.: Thermochim. Acta *63*, 317 (1983); b) Bencivenni, L., Farina, A., Cesaro, S. N., Teghil, R., Spoliti, M.: J. Mol. Struct. *66*, 111 (1980).
185. a) Durig, J. R., Bucy, W. E., Wurrey, C. J.: J. Chem. Phys. *60*, 3293 (1974); b) Durig, J. R., Cooper, P. J., Li, Y. S.: J. Mol. Spectrosc. *57*, 169 (1975).
186. a) Uppal, J. S., Johnson, D. E., Staley, R. H.: J. Am. Chem. Soc. *103*, 508 (1981); see also related studies: b) Kinser, R., Allison, J., Dietz, T. G., de Angelis, M., Ridge, D. P.: J. Am. Chem. Soc. *100*, 2706 (1978).
187. a) Fay, R. C., Lindmark, A. F.: J. Am. Chem. Soc. *105*, 2118 (1983); see also the following related studies; b) Bradley, D. C., Holloway, C.: J. Chem. Soc. A *1969*, 282; c) Harrod, J. F., Taylor, K.: J. Chem. Soc., Chem. Commun. *1971*, 696; d) Bickley, D. G., Serpone, N.: Inorg. Chim. Acta *43*, 185 (1980); e) Byrd, G. D., Burnier, R. C., Freiser, B. S.: J. Am. Chem. Soc. *104*, 3565 (1982).
188. Eysel, H. H., Siebert, H., Groh, G., Berthold, H. J.: Spectrochim. Acta *A26*, 1595 (1970).
189. The relative thermal stability of compounds of the type $(CH_3)_4Ti(Donor)_2$ has been correlated with the electronegativities of the donor atoms: Müller, J., Thiele, K. H.: Z. Anorg. Allg. Chem. *362*, 120 (1968).
190. Francl, M. M., Pietro, W. J., Hout, R. F., Hehre, W. J.: Organometallics *2*, 281 (1983).
191. Lappert, M. F., Pedley, J. B., Sharp, G.: J. Organomet. Chem. *66*, 271 (1974).
192. Abrahamson, H. B., Martin, M. E.: J. Organomet. Chem. *238*, C58 (1982).
193. Chien, J. C. W., Wu, J. C., Rausch, M. D.: J. Am. Chem. Soc. *103*, 1180 (1981).
194. Nesmeyanov, A. N., Nogina, O. V., Fedin, E. I., Dubovitskii, V. A., Kvasov, A., Petrovsky, P. V.: Dokl. Chem. (engl. transl.) *205*, 632 (1972).
195. Druce, P. M., Kingston, B. M., Lappert, M. F., Spalding, T. R., Srivastava, R. C.: J. Chem. Soc. A *1969*, 2106.

3. Chemoselectivity in Reactions of Organotitanium Reagents

temperature and longer reaction times (8–72 h; 50–90% conversion) or higher temperatures [7]:

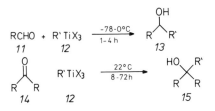

This was a clear signal that titanium reagents should be capable of differentiating between aldehydes and ketones [7]. Another early observation which indicated chemoselective behavior pertains to the reaction of CH_3TiCl_3 (*17*) with bifunctional molecules such as *16* and *19*. In the former case smooth methylation (to be discussed in detail in Chapter 7) occurs at the *tert*-alkyl halide function to the exclusion of C—C bond formation at the ester function [7a–b, 8]. Such selective behavior is not shared by CH_3MgCl [7b]. Furthermore, *17* reacts with both functional groups of *19*, providing *20* and several other products [7]. It was therefore concluded that *17* should add to ketones chemoselectively in the presence of ester groups.

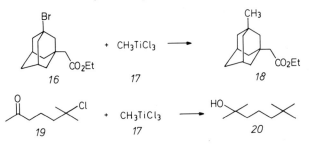

These and other findings [4, 7] led to the working hypothesis that titanation of classical carbanions provides a simple means to control chemo-, regio- and stereoselectivity in reactions with electrophiles [9, 10] (see Chapter 1).

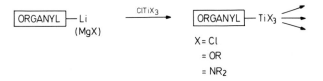

As far as chemoselectivity is concerned, the choice of ligand at titanium is generally not crucial. However, the most versatile ligand system is the isopropoxy group. The titanating agent *23* is readily available by mixing $TiCl_4$ and $Ti(OCHMe_2)_4$ in the right ratio [2, 11]. The large-scale preparation of *23* (e.g., 200 g batches) poses no problems [12].

$$TiCl_4 + 3\ Ti(OCHMe_2)_4 \rightarrow 4\ ClTi(OCHMe_2)_3$$
$$\quad\ 21 \qquad\qquad 22 \qquad\qquad\qquad 23$$

The parent aryl- and alkyltitanium compounds *24* [2, 4, 13] and *25* [2, 4, 5, 14] are accessible in quantitative yield by titanating phenyllithium or methyllithium, respectively. The methyl titanate *25* can be distilled (95% yield) prior to reactions with carbonyl compounds, but an in situ reaction mode is also possible. In fact, the latter is usually the method of choice for most carbanions [10, 15–17]. Grignard reagents are also suitable precursors [6, 7]. For a discussion concerning the stability of common organotitanium compounds, the reader is referred to Chapter 2.

$$\xrightarrow{\text{PhLi}} \text{PhTi(OCHMe}_2)_3$$

23   *24*

$$\xrightarrow{\text{CH}_3\text{Li}} \text{CH}_3\text{Ti(OCHMe}_2)_3$$

25

The first example [4] of a chemoselective reaction of *24* concerns the addition of the reagent (10 mmol) to a mixture of acetyldehyde (*6*; 10 mmol) and acetone (*7*; 10 mmol) at −20 °C. In sharp contrast to PhMgBr [3], a single product *9* was obtained, i.e., the gas chromatogram showed no sign of *10* [4]. Later, another research group provided further examples of aldehyde-selective reactions involving *24* [18]. It should be stressed that these experiments were performed on a synthetic scale, and that they do not provide precise information regarding the relative rates of addition; the results of actual kinetic studies are presented in Chapter 4.

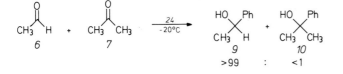

The methyl reagent *25* also turned out to be completely aldehyde-selective [2, 4, 5, 10], e.g.:

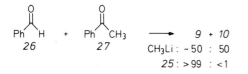

The selectivity of *25* is not restricted to aromatic aldehydes and ketones (e.g., *28/29* → *30*) [4]. Furthermore, the clean reaction of the keto-aldehyde *1* demonstrates that the conclusions derived from intermolecular competition experiments apply fully to polyfunctional molecules [2, 10]. The crude product contains no trace of a ketone adduct, and *32* can be isolated in a yield of 80% [1, 2].

3. Chemoselectivity in Reactions of Organotitanium Reagents

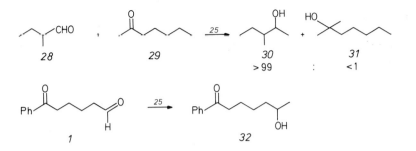

Another simple example involves the addition of ethyllithium (one part) to a mixture of benzaldehyde (one part) and acetophenone (one part), which delivers a ~1:1 mixture of *33* and *34*. Quenching the lithium reagent with ClTi(OCHMe₂)₃ prior to carbonyl addition results in a well behaved species CH₃CH₂Ti(OCHMe₂)₃ which picks out only the aldehyde in excellent yield [10, 12].

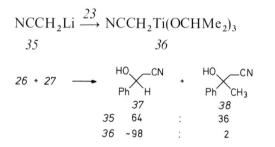

Such control of "carbanion-chemoselectivity" is not restricted to alkyl- and aryl-reagents. Deprotonated nitriles, esters and sulfones as well as other resonance-stabilized and hetero-atom substituted carbanions can also be titanated. Although the precise structure of some of these titanium reagents has not yet been elucidated, they are all temperate in their reactivity, allowing for essentially complete aldehyde-selectivity. An early example is lithiated acetonitrile [2, 9, 15].

$$\text{NCCH}_2\text{Li} \xrightarrow{\;23\;} \text{NCCH}_2\text{Ti}(\text{OCHMe}_2)_3$$

| 35 | | 36 |

26 + 27 ⟶

		37	:	38
35		64	:	36
36	~98		:	2

Further cases [2, 10, 16] refer to such Ti-species as *40* [2, 17] and *41* [1, 2]. Again, the actual structure is currently unknown. The latter species has been tentatively formulated as O-titanated, in analogy to titanium enolates derived from ketones [1].

Chapter 5 contains an H-NMR spectrum of such a species, proving the existence of the Ti—O moiety. Reagent *41* solves the problem of aldehyde-selective aldol addition of ester enolates mentioned in the introductory remarks of this chapter. It should be mentioned that lithium enolates derived from ketones are considerably less reactive, which means that they themselves are aldehyde-selective; in such cases titanation is superfluous [1]. In case of *40*, the C-titanated form should also be considered.

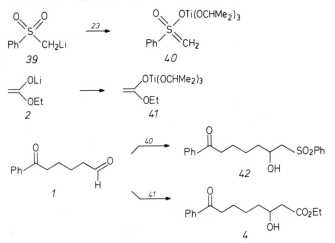

The principles derived from the early studies [4, 9] were later applied to such species as *43–58*. Generally, they were generated from the lithium precursor, and were then reacted in situ with aldehyde/ketone pairs, aldehyde-selectivity being observed in all cases. Here again, the previous cautionary remarks regarding structure and aggregation state apply (see also Chapter 2).

CH₃CH₂CH₂CH₂Ti(OCHMe₂)₃
$$CH_3CH_2CH_2CH_2Ti(OCHMe_2)_3$$

43

44   [2, 9, 12, 18]

45

Me₃SiCH₂Ti(OCHMe₂)₃
$$Me_3SiCH_2Ti(OCHMe_2)_3$$
46   [2, 10, 19]

Ph₂AsCH₂Ti(OCHMe₂)₃
$$Ph_2AsCH_2Ti(OCHMe_2)_3$$
47

Ph₂SbCH₂Ti(OCHMe₂)₃
$$Ph_2SbCH_2Ti(OCHMe_2)_3$$
48   [20]

PhSCH₂Ti(OCHMe₂)₃
$$PhSCH_2Ti(OCHMe_2)_3$$
49

50   [18]
a R = H
b R = CH₃

51

52   [18, 19]
a R = SiMe₃
b R = SO₂Ph

53

54   [10, 21, 22]

3. Chemoselectivity in Reactions of Organotitanium Reagents

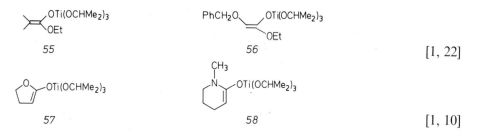

55

56

[1, 22]

57

58

[1, 10]

Although all of the above reagents are aldehyde-selective, the rate of addition varies considerably (Chapter 4). Generally, hetero-atom substituted species, e.g., *46–50*, are least reactive. In some of these cases even aldehydes as reaction partners require room temperature and longer reaction times [2, 18, 20], the yields being only 50–80%. This means that ketones are not likely to react at all. So far, experience has clearly shown that titanium reagents derived from "resonance-stabilized carbanions" react particularly smoothly under mild conditions. These include substituted allyl species such as *52*, cyano- and sulfono-substituted derivatives (*36, 40, 52b, 54*), enolates (*41, 55–56*), enolate equivalents (e. g., *53*) and titanated heterocycles (e.g., *58*). This conclusion also extends to steroselective reactions to be discussed in Chapter 5.

As an example, *55* adds smoothly to acetophenone (−78 °C/4 h; >85% conversion to the aldol). It also reacts aldehyde-selectively (>96:4 product ratio in a competition experiment using benzaldehyde/acetophenone) [1]. The Li-enolate delivers a 58:42 product ratio at −78 °C. A rare case in which titanation does not temper reactivity enough to ensure essentially complete aldehyde-selectivity involves the parent allyl species *60*. At −78 °C it reacts rapidly with a 1:1 mixture of benzaldehyde (*26*) and acetophenone (*27*) to afford an 84:16 mixture of *61* and *62* [10, 19]. However, this problem can be solved by using allyltitanium ate complexes (Section 3.1.2.).

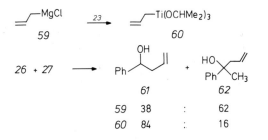

59	38	62
60	84	16

3.1.2 Organotitanium Reagents Bearing Other Ligands

Inspite of the convenience of the *tris*-isopropoxy ligand system, other ligands at titanium were tested, particularly trichlorides (RTiCl₃). The parent compound CH₃TiCl₃ (*17*) was originally generated by the reaction of (CH₃)₂Zn (*63*) with TiCl₄ in CH₂Cl₂ and was shown to behave aldehyde-selectively

[4, 8]. A more convenient procedure involves the reaction of methyllithium with an ether solution of TiCl$_4$, which affords the *bis*-etherate *64* [10]. This reagent displays complete aldehyde-selectivity and excellent conversion (>90%) [22, 23, 24]. Another interesting aspect concerns the smooth addition of *64* to ketones at −20 °C to +22 °C (2–3 h; conversion ~90%), which means that it is considerably more reactive than CH$_3$Ti(OCHMe$_2$)$_3$ (*25*). The main virtue of the reagent has to do with the fact that it adds smoothly to ketones in the presence of additional sensitive functionality (Section 3.5) and to enolizable ketones (Section 3.6). Thus, the advantage over the less reactive CH$_3$Ti(OCHMe$_2$)$_3$ is obvious.

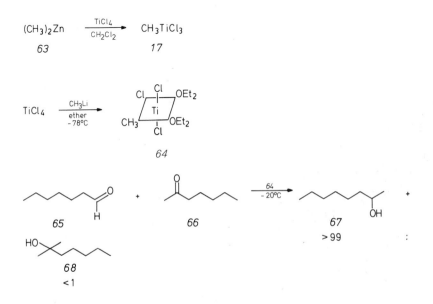

The essential difference between CH$_3$TiCl$_3$ in CH$_2$Cl$_2$ (i.e., *17*) and CH$_3$TiCl$_3$ in ether (i.e., *64*) concerns Lewis acidity of the respective systems. The latter is less Lewis acidic because reagent and initial carbonyl adduct (prior to aqueous workup) are complexed by ether (or THF) [24]. Whereas the CH$_3$Li/TiCl$_4$ method is synthetically quite simple, the mechanism of aldehyde addition is more complicated than in case of CH$_3$Ti(OCHMe$_2$)$_3$ (*25*) (see also Chapter 4). This has to do with the fact that the reacting species need not be the octahedral complex *64* (see Chapter 1 and 2 for a description of such species). It is conceivable that *64*, being essentially coordinatively saturated, must first kick out one or two ether molecules before being capable of adding to carbonyl compounds. Thus, *69* or even *17* may actually be the reactive species, even if they are present to only a small extent. Alternatively, direct addition of *64* to an aldehyde would involve a ligand exchange reaction to form *70* directly prior to aqueous workup. The more stable THF analog of *64* reacts much slower with ketones than does *64*, indicating the necessity of ligand dissociation.

3. Chemoselectivity in Reactions of Organotitanium Reagents

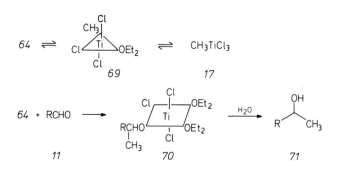

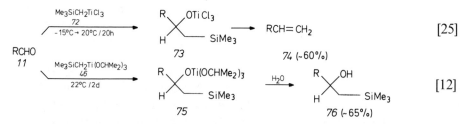

Besides the parent system *17/64*, not too much is known concerning chemoselective addition of RTiCl₃ to aldehydes [10]. This has to do with the fact that in many cases the actual titanation step is not as smooth with TiCl₄ as it is with ClTi(OCHMe₂)₃ [10, 22]. In the former case undesired reduction of Ti(IV) to Ti(III) is more likely [10, 22]. Alkoxy groups are better π-donors than chlorine ligands, (Chapter 2), imparting a greater resistance of titanium toward reduction. The likelyhood of undesired β-hybride elimination of *n*-alkyltitanium species is a related problem, i.e., it is also less likely in case of RTi(OCHMe₂)₃ relative to RTiCl₃ (see also Chapter 2).

Sometimes RTiCl₃ and RTi(OCHMe₂)₃ deliver different products in reactions with carbonyl compounds. An excess of *72* (made by titanating Me₃SiCH₂MgCl in ether) adds to aldehydes (but not to ketones) to form intermediates *73*, which undergo spontaneous Peterson elimination prior to aqueous workup [25]. In contrast, aqueous workup of the reaction of the less reactive *46* with aldehydes *11* leads to the products *76* [2, 12]. If desired, the latter can be converted to *74* using the classical Peterson elimination under basic or acidic conditions, e.g., KH or H⁺/Lewis acids, respectively [26]. The difference in behavior is related to the higher Lewis acidity of *73* relative to *75*. Synthetically, more efficient olefination reagents have been developed, some containing titanium (Chapter 8).

A related case in which RTiCl₃ and RTi(OCHMe₂)₃ afford different products concerns the reaction of PhTiCl₃ (*78*) and PhTi(OCHMe₂)₃ (*24*) with acetaldehyde (*6*) [7d]. The former reagent affords the chloride *82* (64 % isolated) instead of the expected alcohol *9* (which is isolated in 84 % yield in case of *24*). Apparently, the intermediate *80* is S_N1-active, i.e., the carbonium ion *81* is captured by chloride. The overall process is aldehyde-selective; the addition to acetone is fairly slow [7d].

82

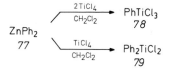

Interestingly, a 1:1 mixture of the (more reactive) dichloride *79* and acetaldehyde (*6*) under similar conditions leads to the alcohol *9* (>90%) [7d]. Other aldehydes react similarly, a process which is completely aldehyde-selective [7d].

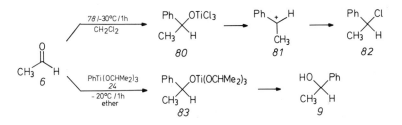

The conversion of the intermediate *80* to *82* is related to the rapid transformation of S_N1-active aryl-activated secondary and tertiary alcohols into the corresponding alkyl chlorides by the action of $TiCl_4$ ($R_3COH \rightarrow R_3CCl$) [15, 22].

Turning to the *tris*-amino ligand system, alkyl and aryl compounds of the type *85* are readily available by titanating Grignard or lithium reagents with $XTi(NEt_2)_3$ (X = Cl [27], Br [28]). As delineated in Chapter 2, they are thermally much more stable than the $RTiCl_3$ or $RTi(OCHMe_2)_3$ counterparts [28].

$$XTi(NEt_2)_3 \xrightarrow[\text{or}]{\underset{R'Li}{R'MgX}} R'Ti(NEt_2)_3$$

$$84 \qquad\qquad\qquad 85$$

a) X = Cl
b) X = Br

a) R' = CH_3
b) R' = $n\text{-}C_4H_9$
c) R' = C_6H_5

Unfortunately, simple alkyl and aryl derivatives *85a–c* fail to afford the Grignard-type adducts in reactions with aldehydes. Instead, low yields (25–60%) of the corresponding amines *88* are observed [10, 15, 29, 30]. This is due to the fact that transfer of the amino group onto the aldehyde is faster than transfer of the carbon nucleophile. The initial adducts *86* then transform into the iminium ions *87* which are captured by the alkyltitanium species. The yields are improved if some $TiCl_4$ is added, which means a one-pot method for geminal amino-alkylation of aldehydes [31].

Fortunately, many synthetically useful carbanions can be titanated with $ClTi(NEt_2)_3$ (*84a*) or $ClTi(NMe_2)_3$ to form titanium reagents which behave "normally" in reactions with carbonyl compounds. The rule for predicting

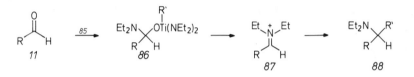

which case pertains in a given situation has been derived empirically [10]: Reactive "resonance-stabilized" species such as *89* [9], *90* [2, 10], *91* [1, 10], *92* [32] and *93* [33] react with aldehydes to form the desired alcohols or aldols in excellent yields 75–95%. Some of them (e.g., *90*, *91* and *92*) have been shown to be aldehyde-selective, while *89* delivers mixtures in aldehyde/ketone competition experiments [29].

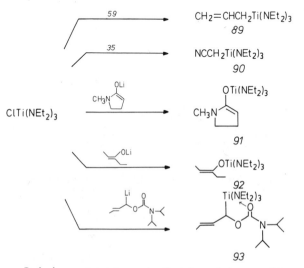

It is important to note that in solving problems of chemoselectivity, the *tris*-amino ligand system offers no advantage relative to the cheaper *tris*-isopropoxy analog RTi(OCHMe$_2$)$_3$ [2, 34, 35a]. However, stereoselectivity depends very much upon the type of ligand, the tris-amino system often delivering the best results [36] (Chapter 5). Thus, it was also necessary to test chemoselectivity [2, 37].

The cyclopentadienyl ligand has also been employed in selective C—C bond formation [10]. The Cp group is such a powerful electron donor ligand (Chapter 2), that reactivity of methyltitanium compounds is sharply reduced. Thus, compounds of the type *94* do not react with aldehydes under normal conditions (22 °C/3 days) [12, 34]. Since it had been known that increasing the number of methyl groups in Ti(IV) compounds increases reactivity toward carbonyl compounds [10, 12], such reagents as *95* were tested. However, even *95* reacts sluggishly with aldehydes [15]. Allyl derivatives (Chapter 5) are more reactive, affording good yields of Grignard-type adducts. As expected, carbonylophilicity increases upon going to *mono*-cyclopentadienyl compounds (e.g., *96*), which behave aldehyde-selectively [12, 22, 35]. Replacing the methyl group in compounds of the type *96* by such C-nucleophiles as allyl or enolate residues results in further increase of reactivity toward aldehydes [34]. It must be remembered that Cp-titanium compounds gain synthetic importance not so much in the area of chemoselectivity, but rather in the area of stereoselective C—C bond formation (Chapter 5).

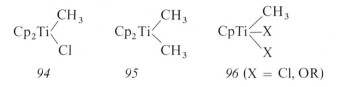

94   95   96 (X = Cl, OR)

3.1.3 Titanium Ate Complexes

Not only the type of ligand systems, but also the number of ligands and the formal charge in Ti(IV) compounds is important in adjusting carbanion-selectivity [10]. The first example to be reported involved titanium enolate ate complexes [32] prepared by adding lithium enolates to $Ti(OCHMe_2)_4$. Thereafter, the allyltitanium ate complex 97 was generated and tested in organic synthesis [10, 29]. Under the conditions used, substitution of an iso-propoxy group to form the neutral compound $CH_2=CHCH_2Ti(OCHMe_2)_3$ (60) does not occur. The ate type of representation is, however, only a formalism, since the aggregation state and coordination number are currently unknown [19]. It is actually unlikely that the reagent is penta-coordinate as shown in 97.

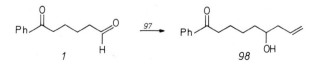

The synthetic significance of 97 is summarized by the following three points:
1) It is easily accessible from cheap reagents;
2) The yields of addition to aldehydes or ketones are essentially quantitative;
3) It behaves chemoselectively, i.e., it distinguishes between aldehydes and ketones and other functionality;
This property is not shared by $CH_2=CHCH_2Ti(OCHME_2)_3$ (60), as previously mentioned (see also Sections 3.2.–3.5. concerning other functionality). Other cases of titanium ate complexes are summarized in Chapter 5. The conclusion regarding aldehyde-selectivity of 97 is based on inter-molecular competition experiments [10, 29], but applies fully to keto-aldehydes. Thus, 1 is attacked only at the aldehyde function (85% isolated yield of 98) [2]. In contrast, 59 delivers a mixture of products.

Attempts to prepare and utilize related alkyl and aryl ate complexes were distinctly less rewarding [10, 15]. Either the interaction of RMgX with *22* turned out to be very slow (if taking place at all), or the ate complexes involving lithium (formed rapidly by addition of RLi to Ti(OCHMe$_2$)$_4$ at −20 °C) were of little synthetic use. For example, *99* actually differentiates between aldehydes and ketones, but yields of products are low (possibly due to competing enolization and/or alkoxy addition) [15]. Nevertheless, the structural features of *99* (which can be obtained in solid form) are of interest and remain to be elucidated.

$$\text{Ti(OCHMe}_2\text{)}_4 \quad \xrightarrow{\text{CH}_3\text{Li}} \quad \text{CH}_3\text{Ti(OCHMe}_2\text{)}_4\text{Li}$$

$$\qquad\quad 22 \qquad\qquad\qquad\qquad\qquad\qquad 99$$

Importantly, the synthetic restrictions regarding alkyl ate complexes of the type *99* do not extent to titanated species generated from resonance-stabilized carbanions [10]. For example, *100* [32], *101* [22], *102* [19] and *103* [17] all add smoothly to aldehydes (conversion >85%) at −40 °C in the presence of ketones. Again, the structure of many of the reagents needs to be studied; for example, *102* and *103* may actually exist in the *N*- or *O*-titanated form, respectively.

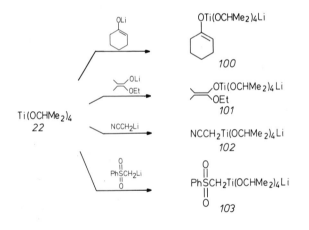

3.2 Aldehyde/Aldehyde and Ketone/Ketone Differentiation

If a molecule contains two (or more) carbonyl groups belonging to the same class of functional groups (e.g., ketone moieties), the problem of site-selectivity arises [10]. Titanium reagents are useful in such situations. Initially, intermolecular competition experiments were performed [9], later di-keto compounds were tested [2, 10]. For example, adding CH$_3$Ti-(OCHMe$_2$)$_3$ (*25*) to a 1:1 mixture of *104* and *105* results in a 92:8 product mixture of *106* and *107*, respectively (conversion 95%). For comparison, CH$_3$MgBr affords a mixture of *106*, *107* and undesired condensation products. In case of the aromatic aldehydes *26* and *108*, an interesting activating effect of the methoxy group is operating [10]. One possible explanation involves complexation of the titanium reagent by the methoxy group. This phenomenon is also effective in case of hydroxy ketones versus the deoxy analogs, as shown in a recent definitive study [38].

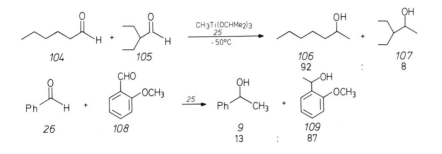

The degree of ketone/ketone discrimination is often substantial, which means that even small differences in steric properties of the carbonyl compounds are "felt" by the titanium reagents [2, 10]. Conversion is >85% in all cases:

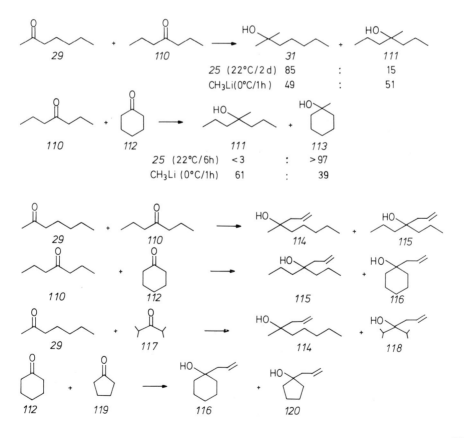

3. Chemoselectivity in Reactions of Organotitanium Reagents

comparison. In all cases addition was performed in ether and conversion was $>80\%$ [15]:

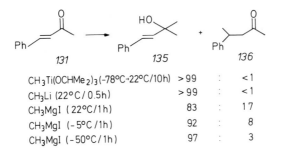

$CH_3Ti(OCHMe_2)_3$ (-78°C-22°C/10h)	>99	:	<1
CH_3Li (22°C/0.5h)	>99	:	<1
CH_3MgI (22°C/1h)	83	:	17
CH_3MgI (-5°C/1h)	92	:	8
CH_3MgI (-50°C/1h)	97	:	3

In order to test whether $CH_3Ti(OCHMe_2)_3$ distinguishes between α,β-unsaturated carbonyl compounds and their saturated analogs, several competition experiments were performed [10, 15]. In case of acyclic carbonyl compounds, the titanium reagent seeks out the α,β-unsaturated substrate preferentially. In case of cyclohexenone/cyclohexanone, the opposite applies, which is difficult to understand off hand. However, in hydride reductions a similar trend has been reported, i.e., cyclohexenone is an exception [42]. This may be due to a combination of electronic and steric effects. Electronically, *131* is less electrophilic than *137*, but the latter is sterically more hindered. In case of *127/112*, electronic effects override steric shielding.

Whatever the actual cause (steric and/or electronic factors), the results clearly demonstrate that titanium reagents are sensitive to structural changes in the acceptor molecule [10, 15].

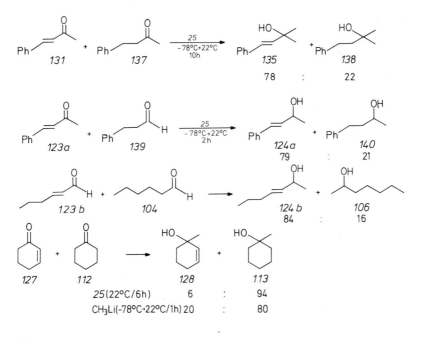

3.4 Aldehyde/Ester and Ketone/Ester Differentiation

Generally, simple alkyltitanium reagents do not undergo C—C bond formation with carboxylic acid esters [4, 8, 10]. In case of CH_3TiCl_3 (*17*) the ester simply forms a Lewis base/Lewis acid adduct [8], much like with $TiCl_4$ [43]. $CH_3Ti(OCHMe_2)_3$ (*25*) shows no interaction with esters; prolonged reaction times lead to transesterification. Only very reactive species such as allyltitanium compounds (e.g., *60*) [19, 44] undergo Grignard-type addition to esters. Based on these and previous results, it can be anticipated that titanium reagents should add chemoselectively to aldehydes in the presence of esters. This is indeed the case [10]. For example, $CH_3Ti(OCHMe_2)_3$ adds completely aldehyde-selectively to a 1:1 mixture of benzaldehyde and ethyl acetate [22]. Generally, classical reagents such as CH_3MgX also show this behavior, so that titanation is unnecessary. However, very reactive reagents such as allylmagnesium chloride do not react as cleanly. In such cases titanation, e.g., with $Ti(OCHMe_2)_4$, produces a well behaved species. Thus, conversion in the following addition is >95 (87% isolated yield of *142*) [2, 19]. It is likely that the $TiCl_4$/allylsilane reagent behaves similarly, although this has not been tested.

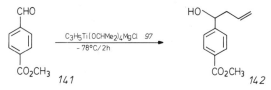

Distinguishing between ketone and ester functions is generally more difficult. Early studies show that titanation leads to reagents which react solely with the ketone function [4, 16a]. Later, this was generalized; sometimes lactonization follows addition [10]. Allyltitanium ate complexes (e.g., *97*) are highly ketone-selective in the presence of esters [2].

An efficient way to obtain β,γ-unsaturated ketones *144* chemoselectively makes use of imidazolides *143* and *89* [44]. Isopropylate *60* or the Grignard reagent *59* cannot be employed because they cause carbinol formation; the same applies to esters in combination with *59*, *60* or *89* [44]. The reaction *143* → *144* does not result in isomerization of the products to the α,β-unsaturated ketones, and is thus to be preferred over other methods.

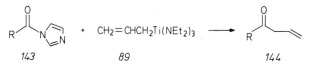

3.5 Reactions in the Presence of Additional Functionality

Several studies have appeared which describe chemoselective additions to aldehydes and ketones in the presence of additional functional groups [2,

10, 16]. In certain cases RMgX is also suitable; however, it was necessary to show that titanium, being a transition metal, performs equally well. In fact, many additional functional groups are tolerated. For example, chloro-alkyl moieties in ketones do not lead to Wurtz-coupling products [2, 10]. Aromatic cyano, nitro and bromo functions also do not interfere with aldehyde addition (e.g., *145a–c* → *146a–c*). Another point of interest con-cerns the reaction of *p*-iodobenzaldehyde (*145d*) with *n*-C$_4$H$_9$Ti(OCHMe$_2$)$_3$ (*43*), which leads to an excellent yield of *148* [18], a process which fails com-pletely with *n*-butyllithium (probably due to undesired halogen/lithium exchange!). Also, in case of *153* and *157* classical reagents such as CH$_3$Li or CH$_3$MgX lead to poor yields of addition products (~30–40%) [2].

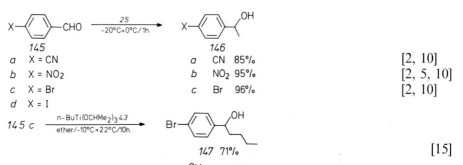

145		**146**		
a	X = CN	*a*	CN 85%	[2, 10]
b	X = NO$_2$	*b*	NO$_2$ 95%	[2, 5, 10]
c	X = Br	*c*	Br 96%	[2, 10]
d	X = I			

147 71% [15]

148 82% [18]

149 → **150 89%** [2]

151 → **152 76%** [15]

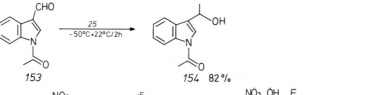

153 → **154 82%** [2, 10]

155 → **156 98%** [18]

157 → **158 81%** [2, 10]

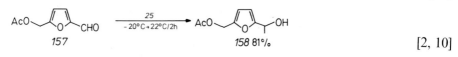

It was important to know whether more reactive methyltitanium species (which add more efficiently to ketones) tolerate additional functionality. To this end, two systems were studied: $(CH_3)_2Ti(OCHMe_2)_2$ (*160*) and the previously decribed $CH_3Li/TiCl_4$/ether reagent (Section 3.1). The dimethyl reagent is prepared by treating *159* with two equivalents of CH_3Li. It turned out to be considerably more reactive towards ketones than the monomethyl reagent $CH_3Ti(OCHMe_2)_3$ (*25*). By choosing the amount of reagent *160* in such a way that only one active methyl group is transferred onto ketones, excellent yields of tertiary alcohols are obtained under mild conditions ($-30\,°C \rightarrow 22\,°C/2$–4 h) [2, 9, 12]. The increased reactivity of *160* relative to *25* has to do with the greater driving force of titanium to form a new Ti—O bond (Chapter 2). The former reagent has only two such bonds to start with, the latter three. As more Ti—O moieties are introduced at the expense of Ti—C units, Lewis acidity is also reduced, and bulkiness increases. In line with these conclusions is the observation that upon going from *160* to higher methylated titanium reagents, carbonylophilicity increases even more (Chapter 4). Despite the increased rate of ketone addition, *160* tolerates additional functionality such as cyano groups, e.g., *161* $\rightarrow$ *162* (single diastereomer) [2, 10]. Also, *160* is well suited for addition reactions with highly enolizable ketones [9] (Section 3.6). Of course, the reagent transfers both methyl groups onto aldehydes (if a 1:2 ratio is chosen) [12, 22], the intermediate *163* being related to the parent compound $CH_3Ti(OCHMe_2)_3$ (*25*).

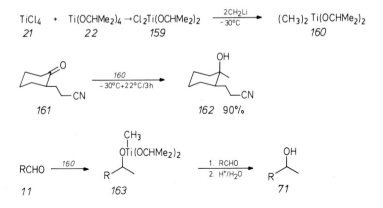

Another advantage of *160* over *25* concerns addition to certain nitroketones [10]. It is known that most nitro compounds react with Grignard reagents to form mixtures of various products [45]. Thus, it was of interest to see whether the reaction of *164* with methyltitanium compounds proceeds chemoselectively [2, 10]. Whereas the monomethyl reagent *25* affords <20% of the desired adduct *165* after two days at room temperature, the more reactive reagent *160* leads to 78% of *165* after a reaction time of 3 hours at room temperature [2, 10]. It is possible that the major reaction of *25* is undesired titanation of *164* at the acidic methylene moiety, forming titanium nitronates [12].

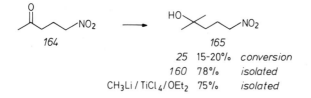

25	15–20%	conversion
160	78%	isolated
CH₃Li / TiCl₄ / OEt₂	75%	isolated

As previously pointed out, the addition of methyllithium to TiCl₄ in ether results in the bis-etherate of CH₃TiCl₃, a reagent which adds very efficiently to aldehydes (−78 °C/1 h) and ketones (−20 °C → 22 °C/2–3 h), the process being completely aldehyde-selective in relevant cases [23, 24]. It was therefore worthwhile to test additional functionality. Indeed, nitro, cyano, ester groups as well as primary alkyl halide entities are tolerated [23]. For example, in case of the nitro-ketone *164*, a 75% yield of *165* was obtained (−20 °C → +22 °C/2 h) [23]. Thus, CH₃Li/TiCl₄/ether is a fairly non-basic, highly chemoselective and easily accessible reagent. It also adds Grignard-like to *p*-nitroacetophenone [23], whereas CH₃MgX attacks the aromatic ring [45c].

These aspects of titanium chemistry have been applied in more complex situations, e.g., in the synthesis of macrocyclic lactones such as (±)phora-cantholide [46]. The trifunctional molecule *168* was treated with (CH₃)₂-Ti(OCHMe₂)₂ (*160*), affording 85% of *172* as a single diastereomer. Whereas the cause of this interesting diastereoselectivity has not been completely elucidated [46, 47], the result suggests that an enantioselective synthesis of phoracantholide should be possible if the precursor *168* could be obtained in optically active form. It is of interest to note that the yield of addition product using CH₃MgI is much lower and that a mixture of diastereomers is obtained. In case of CH₃Ti(OCHMe₂)₃ (*25*), the yield is acceptable (60%), but a diastereomeric mixture results [48].

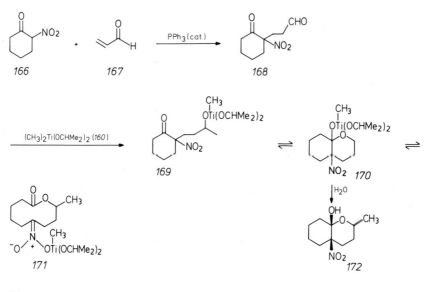

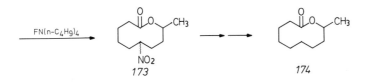

FN(n-C₄H₉)₄

173   174

This simple methodology has been applied to the synthesis of other macrocyclic lactones, e.g., (±)-dihydrorecifeiolide (*175*) and (±)-15-hexa-decanolide (*176*) [46, 49].

175   176

In related work, chemoselective addition of allyl groups to *177* was studied [50]. Whereas the allyltitanium reagents *60* and *97* delivered poor yields of the desired adduct *178*, the combination TiCl$_4$/CH$_2$=CHCH$_2$SiMe$_3$ [40, 51] resulted in quantitative conversion [50]. The poor performance of *60* and *97* surprises. However, this may be due to the reaction mode; the keto-aldehyde *177* was slowly added to a solution of the allyltitanium reagents. In case of polyfunctional molecules, the order of addition should be the opposite, i.e., titanium reagents should be added to the organic substrate (see previous discussion).

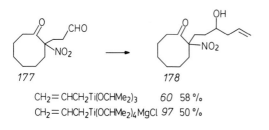

177   178

CH₂=CHCH₂Ti(OCHMe₂)₃   *60* 58 %
CH₂=CHCH₂Ti(OCHMe₂)₄MgCl *97* 50 %

A novel type of chemoselectivity was observed in the attempted *mono*-addition of methylmetal reagents to di-aldehydes such as *179* [10, 15]. Whereas CH₃MgI results in a mixture of *mono*- and *bis*-adduct (*181/182*) as well as starting material *179*, *25* induces clean *mono*-addition to form *181* (72% isolated). The bis-adduct *182* is accessible in 90% yield by using two equivalents of the titanium reagent [52] (diastereoselectivity is discussed in Chapter 5). The reason mono-addition can be controlled relates to the intermediacy of the primary adduct *183* which is in equilibrium with the titanated hemi-acetal *184*. The rate constants k$_1$, k$_2$, k$_3$ and k$_4$ are such that the initial process, governed by k$_1$, is most rapid. Such intramolecular in situ protection (*184*) also occurs in case of other related di-aldehydes, which underlines the generality of the method [22]. Furthermore, such functionalities as RBr, RCO₂Et, RCN and ArNO₂ are tolerated [22].

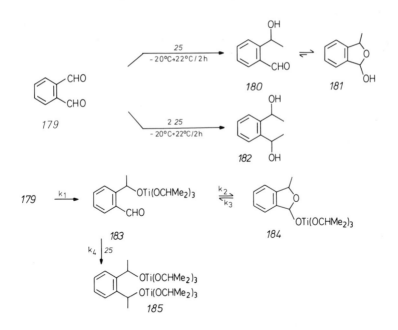

Finally, it has been demonstrated that alcoholic functions are tolerated to some extent in aldehyde addition reactions [53]. A 1:1 mixture of benzaldehyde and isopropanol reacts with one equivalent of CH₃Ti-(OCHMe₂)₃ (25) at −25 °C in THF to yield 55% of 2-phenylethanol [2]. This means that protonation of the titanium reagent (to form methane) occurs at about the same rate as aldehyde addition [2]. Surprisingly, even CH₃TiCl₃ (17) shows this behavior [54]; however, chromium reagents are better suited [54] (Section 3.8).

[54]

3.6 Addition to Enolizable Ketones

A well known drawback of classical reagents RLi and RMgX concerns their tendency to deprotonate enolizable ketones, a side reaction which is sometimes a major problem [39, 55]. Since titanium compounds are much less basic than RLi or RMgX (Chapter 2) but still carbonylophilic, they are generally well suited in such situations. The first example to be reported concerns the conversion 186 → 187 [9]. Whereas CH₃Li affords less than 55% of 187 due to competing enolization, the dimethyltitanium reagent 160 results in clean addition [9]. Diastereoselectivity is also better than in case of CH₃Li (Chapter 5). Interestingly, 25 is not well suited for addition, enolization occurring to ~50% [12]. In fact, a case in which it performs distinctly less well than CH$_3$Li has been reported [52, 56]. α-Tetralone (*188*) reacts with CH$_3$Li to form 85% of *189* [10], whereas *25* results in ~50% enolization [10, 52, 56].

The dimethyltitanium compound *160* [52] or *mono*-methylzirconium reagents *190* [52, 56] result in clean addition. In view of the efficient reaction of CH$_3$Li [10], the latter processes are only of mechanistic interest. Zirconium reagents *190* are clearly better suited than *25*, but the conclusion that this is due to the higher basicity of titanium reagents [56] is not entirely satisfactory. It focuses only on one aspect (deprotonation), where in fact addition must also be considered. Certainly, CH$_3$Li is more basic than *25*, and yet poses no enolization problem in this case.

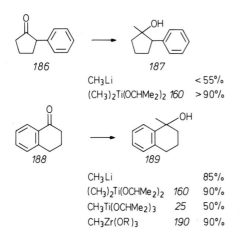

186	187	
CH$_3$Li		<55%
(CH$_3$)$_2$Ti(OCHMe$_2$)$_2$ *160*		>90%

188	189	
CH$_3$Li		85%
(CH$_3$)$_2$Ti(OCHMe$_2$)$_2$ *160*		90%
CH$_3$Ti(OCHMe$_2$)$_3$ *25*		50%
CH$_3$Zr(OR)$_3$ *190*		90%

A more realistic test is provided by the easily enolizable β-tetralone (*190*). In this case it is clear that CH$_3$Li/TiCl$_4$/ether is the method of choice [23].

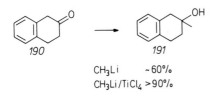

190	191	
CH$_3$Li		~60%
CH$_3$Li/TiCl$_4$		>90%

The sterically hindered and enolizable ketone *192* fails to undergo C—C bond formation with CH$_3$MgX or CH$_3$Li to any appreciable extent [52]. Whereas *25* and *160* are not much better, Ti(CH$_3$)$_4$ (*195*) and Zr(CH$_3$)$_4$ (*196*) are surprisingly well suited [52]. Thus, the zirconium reagent (*196*) is a highly reactive reagent of low basicity [10, 52], to be employed in extreme situations. The ratio of *196* to ketone should be 1:1, which means that only one active methyl groups is utilized. Of course, in case of simple aldehydes and ketones, all four methyl groups can be made to react. However, reactivity toward carbonyl compounds decreases with decreasing number of methyl groups attached to zirconium (or titanium; see Chapter 4), which means more enolization in cases like *192*. The $CH_3Li/TiCl_4$/ether system results in $< 10\%$ of *193* [23]. The super methylating power of *196* has been demonstrated in the conversion *197* → *198*, which cannot be realized using CH_3Li [57].

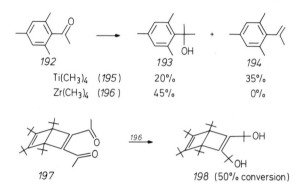

	192	193	194
Ti(CH₃)₄	(*195*)	20%	35%
Zr(CH₃)₄	(*196*)	45%	0%

197 → [196] → *198* (50% conversion)

3.7 Limitations of Organotitanium Reagents

Several limitations of titanium reagents in selective addition reactions to carbonyl compounds have been uncovered [10]. As delineated in Chapter 2, secondary and tertiary alkyltitanium compounds are generally unstable due to β-hydride elimination. Thus, reacting isopropyl- or *tert*-butylmagnesium chloride with $ClTi(OCHMe_2)_3$ (*23*) followed by the addition of aldehydes fails to give appreciable yields of Grignard adducts [10], reduction to primary alcohols as well as pinacol formation setting in. Alkyllithium reagents behave similarly [10, 22]. The trouble may occur at two different stages;
1) Attempted titanation of branched RMgX or RLi results in direct reduction of $ClTi(OCHMe_2)_3$, which sets the stage for reduction and low valent titanium mediated reductive dimerization of the aldehyde subsequently added;
2) the desired species $RTi(OCHMe_2)_3$ may actually be formed to some extent, but then reduces directly the aldehyde via β-hydride transfer (analogous to the well-known Grignard reduction [55, 58]).
    Experiments designed to illuminate these processes have not been carried out. Although branched alkyltitanium compounds with amino ligands are thermally stable (Chapter 2), they cannot be used for Grignard-type additions due to competing transfer of amino groups (Section 3.11) [10].
    Another type of limitation concerns vinyltitanium reagents. It has been reported that attempts to titanate *199* and to react the species *200* with aldehydes fails to afford the usual adducts [56]. Instead, *200* undergoes oxidative dimerization to *201* at low temperatures. However, it is presently not clear whether this phenomenon is general for all vinyltitanium compounds, particularly in view of the fact that some of them have been isolated [59]. In fact, preliminary experiments with $CH_2=CHTi(OCHMe_2)_3$ show that addition to aldehydes is feasible [22].

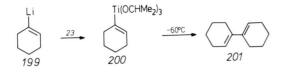

Titanated acetylenes have also not been studied systematically. Initial experiments seem to indicate that the reactions are not as clean as in case of *n*-alkyl- or allyltitanium reagents [22]. Recently, examples of stereoselective additions to chiral aldehydes using $RC\equiv CTiX_3$ have been reported to proceed with moderate yields [60] (Chapter 5). It is currently unclear whether competing oxidative dimerization occurs or whether the actual carbonyl addition is sluggish (or both).

Sometimes titanation tempers "carbanion-reactivity" to such an extent that no addition to carbonyl compounds occurs. A case in point is the titanated form *204* of DMSO, which fails to add to benzaldehyde under a variety of conditions (e.g., $-20\,°C \rightarrow 22\,°C/19$ h) [17].

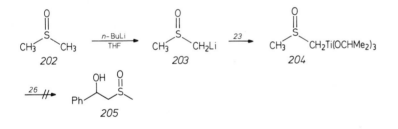

The reason for this unexpected failure (α-titanated sulfones react efficiently [2]) is currently not understood [17, 61]. In fact, initial H-NMR studies of *204* do not reveal whether the species is C-titanated as shown or whether it is O-titanated (*206*) [17]. Furthermore, the aggregation state is unknown. Perhaps the lack of carbonylophilicity is caused by intramolecular complexation as shown in *207/208* [17], or by related intermolecular bridging leading to aggregation. This could reduce Lewis acidity of titanium [17]. Parenthetically, the structure of *203* has also not been elucidated [62].

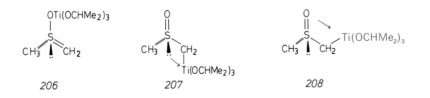

3.8 Hints on How to Use Organotitanium Compounds

Several additional aspects should be kept in mind when applying organotitanium chemistry as described here. If carbanions are to be titanated, alkoxy, amino and chloro ligands are most likely to ensure success; sulfur or phosphorus ligands have not yet been tested. Sometimes reduction to low valent titanium (e.g., purple Ti(III) species) competes with the desired titanation. This may occur even if no β-hydrogen atoms are present in the reagent, i.e., in cases where electron rich carbanions undergo electron-transfer onto the titanating agent. Such processes, although rare, are most likely to occur with $TiCl_4$. The ease of reduction decreases in the series $TiCl_4 < ClTi(OCHMe_2)_3 < ClTi(NEt_2)_3$; as the ligands become better π-donors (Chapter 2), electron density at titanium increases, making it less susceptible to reduction [10]. Addition of amines or pyridine prior to titanation has a similar effect [10, 22].

Concerning workup, simple quenching with ice-water or dilute HCl-solution usually poses no problems (in case of $RTiCl_3$, cold water is the method of choice). In contrast, if Na_2CO_2 is used, emulsions containing TiO_2 are likely to be formed which hamper workup. If this occurs (even in case of acidic workup), saturated aqueous solutions of NH_4F (or KF) should be employed. This means basic conditions; the fluoride ions de-titanate the adduct due to the formation of strong Ti—F bonds (Chapter 2). Whatever the exact structure of the cleaved Ti-species, they are fairly soluble in water and have a certain lifetime before hydrolyzing to insoluble TiO_2 [10, 22].

3.9 Why Does Titanation of Carbanions Increase Chemoselectivity?

The rate of carbonyl addition of $CH_3Ti(OCHMe_2)_3$ (25) is considerably lower than that of CH_3Li or CH_3MgX (Chapter 4). Generally, the faster a reaction, the lower the selectivity. However, this does not explain the increased chemoselectivity of $RTiX_3$ relative to the above classical reagents, since the cause of the lower rate of addition remains unclear. This question is difficult and cannot be answered satisfactorily at the present time. Several factors can be speculatively singled out. It is likely that the C—Ti bond is considerably less polar than the C—Li or C—MgX analogs (Chapter 2), resulting in a lower rate of carbonyl addition and consequently greater selectivity. Also, the three ligands at titanium in $RTiX_3$ are rather bulky (e.g., isopropoxy groups). Thus, as the reagent adds to the carbonyl site, steric interaction between the ligands and the substituents on the substrate is certain to be operating. It is important to note that the Ti—O bond is fairly short (1.7–1.9 Å), which means that in the transition state of the addition to carbonyl compounds steric repulsion is greater than in case of other CH_3-metal reagents. This may well be the reason why zirconium reagents are not so chemoselective (Zr—O = 2.1 Å). Titanium reagents respond effectively to small steric changes, which makes discrimination between two sterically different sites possible, e.g., in case of two ketones. This steric factor is also expected to be important in aldehyde/ketone differentiation, in addition to the fact that aldehydes are better electrophiles than ketones. The intricacies of carbonyl addition are just beginning to be unravelled (Chapter 4).

3.10 Comparison with Other Organometallic Reagents

Previous to the research on titanation, such indiscriminate reagents as RMgX and RLi had been converted into zinc or cadmium analogs in order to perform ketone syntheses from carboxylic acid chloride [63]. However, RZnX and RCdX (or the dialkyl compounds) generally do not react smoothly with aldehydes or ketones, so that adjustment of chemoselectivity using these metals is not feasible. Manganese reagents likewise convert carboxylic acid chlorides into ketones [64]. Also, addition to aldehydes is faster than to ketones, although differentiation is not complete [65] (Table 2).

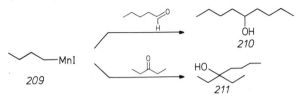

Table 2. Relative Rates of Addition of *209* to Carbonyl Compounds [65]

Temp. (°C)	% Conversion after 15 min.	Product
+20	89	*210*
+20	89	*211*
−30	85	*210*
−30	30	*211*
−50	72	*210*
−50	10	*211*

Cuprates appear to add to aldehydes faster than to ketones [66], although mixtures are obtained [67]. Another disadvantage concerns the loss of one active alkyl group. The following data are typical [67].

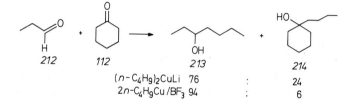

$(n\text{-}C_4H_9)_2CuLi$	76	:	24
$2n\text{-}C_4H_9Cu/BF_3$	94	:	6

3. Chemoselectivity in Reactions of Organotitanium Reagents

In competition experiments employing aldehyde/ketone pairs, the BF_3-mediated addition of allyltri-n-butyltin turned out to be only slightly aldehyde selective ($\sim 70:30$ product mixtures) [68]. In contrast, the reagent system $TiCl_4/CH_2=CHCH_2-SiMe_3$ appears to be chemoselective [50] (see Section (3.4), although the generality remains to be established. Allylboron reagents add readily to aldehydes, but sluggishly (or not at all) to ketones, a property which makes chemoselective transfer possible [69]. The silicon, tin and boron systems are restricted to allyl groups; alkyl derivatives do not add to aldehydes.

Organozirconium reagents are aldehyde-selective, although not always 100% [22, 52, 56]. Thus, the Ti-analogs are to be preferred (they are also cheaper). However, highly branched alkyltitanium reagents are unstable (Chapter 2), in contrast to the Zr-analogs. In such cases zirconium chemistry is the method of choice [56].

Recently, interesting studies involving methyl transition metal compounds (Cr, Mo, Ta, Nb) have appeared [54, 70, 71]. Although an excess of reagent is usually required (e.g., *215*) [71], competition experiments point to complete aldehyde selectivity. In fact, most of the reagents, e.g., *215*, fail to react with ketones, as shown by separate experiments [71]. The chromous chloride induced allylation is quite useful and has been applied in stereoselective additions [70, 72].

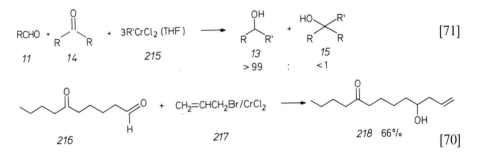

Of synthetic interest are also Grignard-type additions and olefinations in the presence of protic functionality [54]. Although certain titanium reagents add to aldehydes in the presence of alcoholic HO-functions (Section 3.5), chromium reagents are more efficient. This shows that the search for new organometallic reagents can be rewarding.

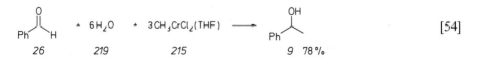

Recently, the problem of Grignard-type addition to enolizable ketones (Section 3.6) has been re-considered using organocerium reagents [73]. A variety of alkyl-, alkenyl- and alkynylcerium derivatives were added cleanly to enolizable ketones, e.g., *190 + 221 → 222*, demonstrating superior behavior relative to titanium analogs. Perhaps in case of the parent compound CH_3CeCl_2, the more readily available $CH_3Li/TiCl_4$/ether reagent [23] (Section 3.6) is to be preferred. Organozirconium reagents behave similarly [52, 56].

$$n-C_4H_9Li \xrightarrow{\text{CeCl}_3} n-C_4H_9CeCl_2 \xrightarrow{190}$$

[73]

220   221   222   88%

In summary, organotitanium(IV) chemistry is complementary to other organometallic systems. For example, cuprates add in a 1,4-manner to α,β-unsaturated carbonyl compounds, while titanium reagents react 1,2-regio-selectively and chemoselectively in the presence of other functionality. Concerning the increase in chemoselectivity of carbanions via transmetallation, titanium is the most versatile metal to date. Advantages include high yields, low cost, ease of performance and the fact that no toxic materials are formed upon workup (ultimately TiO_2). The possibility of varying the nature of the ligand at titanium is also noteworthy (particularly in controlling stereoselectivity as described in Chapter 5). In case of failures, e.g., branched alkyltitanium compounds, other metal systems such as zirconium or cerium fill the synthetic gap. Finally, important synthetic transformations such as oxidative additions, cyclodimerization of dienes, C—H activation and certain substitution reactions are best performed using other transition metals [74]. Lanthanide reagents are beginning to be applied to organic synthesis [75]. Reactions include Grignard-type additions to ketones using $R-Br/SmI_2$, chemoselective reduction of aldehydes and ketones as well as selective pinacol coupling.

3.11 Reversal of Chemoselectivity: Chemoselective in situ Protection of Carbonyl Compounds

It would be of synthetic interest to perform C—C bond formation chemoselectively at less reactive carbonyl sites, e.g., at the ketone function of a keto-aldehyde or at the sterically more hindered site of a di-ketone. Such a reactivity pattern would mean reversal of the chemoselectivity as previously described for organotitanium reagents (Sections 3.1–3.4). The first indication that such processes are in fact possible involves the anomalous behavior of the amino-ate complex 224 [29]. In contrast to the alkoxy analog $CH_2=CHCH_2Ti(OCHMe_2)_4MgCl$ (97), it adds chemoselectively to ketones in the presence of aldehydes.

Thus, simply switching from alkoxy to amino ligands reverses chemoselectivity [10, 29]! For example, in case of 65/29 aqueous workup affords essentially only 114 besides non-reacted aldehyde 65 [29]. Similar observations were made using other aldehyde/ketone systems.

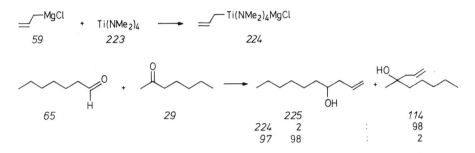

Completely opposite chemoselectivity was also observed in going from *226* to *227* [29].

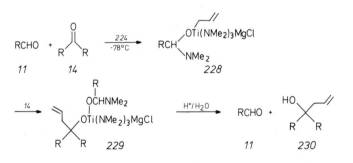

The reason for the above "anomaly" becomes apparent when realizing that the amino reagents *224* and *227* have two reactive ligands: the carbon and the nitrogen nucleophile. Since the organic substrate has two potential acceptor sites (aldehyde or ketone), four different processes are possible in the initial step. Of these, chemoselective transfer of the amino group onto the aldehyde is rapid, leading to the protected form *228*. This leaves the ketone function untouched, which can then react with another amino group in a similar manner, or undergo C—C bond formation via transfer of the carbon nucleophile. The latter is faster, forming the intermediate *229*. Aqueous workup then affords the ketone adduct *230* in excellent yield (~95% conversion) and also regenerates the aldehyde *11*.

Obviously, the underlying factor involves chemoselective in situ protection of aldehydes [29]. Lithium ate complexes such as $CH_2=CHCH_2Ti(NMe_2)_4Li$ behave similarly, although the degree of chemoselectivity is not 100% [10, 19]. Unfortunately, the phenomenon does not extend to simple alkyl ate complexes, e.g., $CH_3Ti(NMe_2)_4Li$, the actual yields of addition products being poor [19, 22]. However, if the same components used in making the amino ate complexes are employed in a different manner, reversal of chemoselectivity is in fact possible in a one-pot procedure [19].

This interesting goal is reached by using compounds *223* or *232* [37]. The latter has been synthesized on a large scale using readily available materials [27a], i.e., lithium diethylamide is made by the Ziegler procedure (HNEt₂|Li|-styrene) followed by titanation. Since the Ti—O bond is thermodynamically strong (Chapter 2), addition to carbonyl compounds is expected to be exothermic. Indeed, *223* and *232* both add rapidly to aldehydes at −78 °C to form intermediates *233* [37]. Transfer of another amino group is possible, but occurs at a slightly lower rate. Such adducts have been characterized by ¹H-NMR spectroscopy at low temperatures [19]. At temperatures above −25 °C they begin to decompose to enamines [76]. Importantly, the reaction of ketones with *223* is slow at −78 °C, but smooth at −50 °C to −30 °C. This clearly suggests that chemoselective transfer onto aldehydes should be possible. Also, the fact that the more bulky reagent *232* fails to add to ketones at temperatures below −30 °C (at −10 °C to 22 °C enamine formation sets in), shows that the process is extremely sensitive to steric factors.

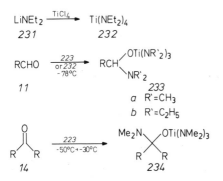

The strategy of a one-pot procedure for reversal of chemoselectivity is clear: in situ protection followed by addition of classical carbanions to the non-protected and thus less reactive carbonyl group. This works well, provided the carbanion reaction proceeds smoothly at temperatures below −25 °C [37] (e.g., *29* → *236*). Also, *1* undergoes aldol addition solely at the ketone moiety (79% of *3* isolated) [1, 37].

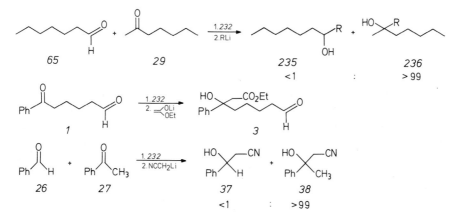

Differentiation between two different ketone sites is also possible, leading to C—C bond formation at the sterically more hindered site [37]. For example, the addition of *223* (one part) to *121* (one part in THF at −50 °C/1 h) followed by treatment with CH_2=$CHCH_2MgCl$ (−50 °C → −40 °C/2 h) and acidic workup (dil. HCl) results in clean formation of *237* as a single diastereomer (95% isolated). This may be contrasted to reactions in the absence of $Ti(NMe_2)_4$: The Grignard reagent itself affords a mixture of adducts, while CH_2=$CHCH_2Ti(OCHMe_2)_4MgCl$ reacts solely at the sterically less hindered C^3-position [37] (Section 3.2). The lithium enolate of ethyl ester can also be made to react at the more hindered C^{17}-position of *121* following in situ protection using *223*. The aldol adduct is formed 100% chemo- and stereoselectively (α-attack) and can be isolated with 80% yield [77].

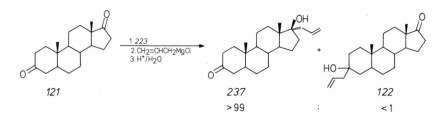

| 121 | | 237 | | 122 |
| | | > 99 | : | < 1 |

Finally, addition to esters in the presence of aldehydes poses no problems. In the following case the aldehyde adduct was not detected in the crude product (72% isolated yield of *238* [19]).

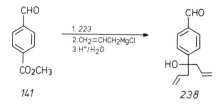

141   238

Some of these transformations can be performed using classical protective methods [78]. However, they involve three steps: Protection (which is not always chemoselective), reaction and deprotection. Thus, in situ methods are to be preferred [79], when possible. Since less reactive carbanions require temperatures of −10 °C to 0 °C for smooth addition, the present in situ protective method fails; the protected forms *233/234* decompose above −25 °C. Therefore, other metal amides were tested. However, zirconium, boron and zinc derivatives turned out to be less efficient than the titanium reagents [19]. In contrast, preliminary results using the manganese amides *240* and *241* are promising, since the analogous carbonyl adducts are stable up to at least 0 °C [77]. Furthermore, both amino groups transfer readily onto aldehydes or ketones, the process being aldehyde-selective in relevant cases. However, more complicated systems such as the di-ketone *121* need to be studied before a final evaluation regarding the relative merits of various metal amides can be made.

A final point concerns attempted chemoselective reduction of more hindered ketones using titanium amide mediated protection followed by reduction with such reagents as diisobutylaluminum hydride (DIBAL) [1]. The methodology fails because the protected ketones compete for the DIBAL, resulting in reductive amination. In the absence of additional carbonyl groups, this reaction can be optimized by using 0.5 equivalents of *232* [1, 22]. Similar reactions with *240* or *241* are not as smooth.

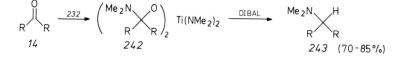

3.12 Organotitanium Reagents from Non-Organometallic Precursors

Besides titanation of carbanions, organotitanium reagents can also be prepared by the addition of Ti(OR)$_4$ or TiCl$_4$ to non-organometallic precursors. Several examples of such a strategy are currently known. In a Reformatsky-type addition, a mixture of Ti(OCHMe$_2$)$_4$ and an aldehyde or ketone *244* (R^1, R^2 = H, alkyl, aryl) in ether was treated with gaseous ketene and the products *245* isolated in good yields [80]. Although the mechanism of this reaction has not been elucidated, an organotitanium intermediate *246* (or the O-titanated form *247*) was postulated [80]. These are related to the intermediates *41* made by titanation of the lithium enolate of ethyl acetate [1, 37]. It is not clear whether the method can be generalized to include substituted ketenes, and whether other titanating agents such as TiCl$_4$ and Ti(NR$_2$)$_4$ can also be used.

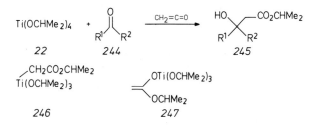

3. Chemoselectivity in Reactions of Organotitanium Reagents

In a different approach, isonitriles were used as precursors. It has been known for some time that they react with TiCl₄ to form the α,α-adducts [81]. However, the organic chemistry of these interesting species was not studied until recently [82]. The α,α-adducts (e.g., *249*) add smoothly to aldehydes or ketones, forming N-methyl-α-hydroxycarboxamides *250* following aqueous workup. Thus, the overal process is related to the classical Passarini reaction [83]. Compounds of the type *250* can also be prepared by the reaction of certain isonitriles and an excess of a carbonyl component in the presence of mineral acids [84]. It is likely that TiCl₄ will also add to other reactive π-systems such as allenes and other cumulenes as well as diazo compounds.

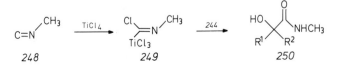

248   249   250

Finally, O-silyl cyclopropanone ketals *251* react with TiCl₄ to form the titanium homoenolates *252*, which add to aldehydes [85]. Evidence for the proposed structure *252* was derived from NMR experiments. *252c* is a deep purple, moderately air-sensitive compound that is stable at room temperature for several months. It is dimeric, as shown by cryoscopy. Since addition of *252* to ketones is sluggish, the reagents are highly aldehyde-selective in relevant cases.

251   252   253

a R = CH₃
b R = CH₂CH₃
c R = CHMe₂

Recently, an important improvement has been achieved by treating *252c* with Ti(OCHMe₂)₄ which generates the dichloroisopropoxy analog [85b]. This reagent is more reactive, adding to ketones and sterically hindered aldehydes. It has been applied in an elegant synthesis of depresosterol [85b].

In case of donor-acceptor-substituted cyclopropanes such as *254*, reactivity is greater so that even ketones react [86]. The products are either lactols *256* or olefins *257*, depending upon the mode of workup. The former are easily transformed into synthetically useful 2,3-dihydrofurans. Besides *254* several other derivatives have been described [86].

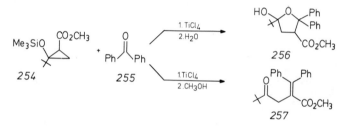

254   255   256

257

References

1. Peter, R.: Dissertation, Univ. Marburg 1983.
2. Reetz, M. T., Westermann, J., Steinbach, R., Wenderoth, B., Peter, R., Ostarek, R., Maus, S.: Chem. Ber. *118*, 1421 (1985).
3. Kharasch, M. S., Cooper, J. H.: J. Org. Chem. *10*, 46 (1945).
4. Reetz, M. T., Steinbach, R., Westermann, J., Peter, R.: Angew. Chem. *92*, 1044 (1980); Angew. Chem., Int. Ed. Engl. *19*, 1011 (1980).
5. Weidmann, B., Seebach, D.: Helv. Chim. Acta *63*, 2451 (1980).
6. Reviews of common organotitanium compounds: a) Segnitz, A.: in Houben-Weyl-Müller, "Methoden der Organischen Chemie", Vol. 13/7, p. 263, Thieme Verlag, Stuttgart 1975; b) Gmelin Handb., Titanorganische Verbindungen, Vol. 40, N.Y. 1977; c) Wailes, P. C., Coutts, R. S. P., Weigold, H.: "Organometallic Chemistry of Titanium, Zirconium and Hafnium", Academic Press, N.Y. 1974; d) see also ref. [16].
7. a) Westermann, J.: Diplomarbeit, Univ. Bonn 1980; b) Steinbach, R.: Diplomarbeit, Univ. Bonn 1980; c) Wenderoth, B.: Diplomarbeit, Univ. Marburg 1980; d) Peter, R.: Diplomarbeit, Univ. Marburg 1980.
8. Reetz, M. T., Westermann, J., Steinbach, R.: Angew. Chem. *92*, 933 (1980); Angew. Chem., Int. Ed. Engl. *19*, 901 (1980).
9. Reetz, M. T., Steinbach, R., Wenderoth, B., Westermann, J.: Chem. Ind. *1981*, 541.
10. Review: Reetz, M. T.: Top. Curr. Chem. *106*, 1 (1982).
11. Dijkgraaf, C., Rousseau, J. P. G.: Spectrochim. Acta *24A*, 1213 (1968).
12. Westermann, J.: Dissertation, Univ. Marburg 1982.
13. Holloway, H.: Chem. Ind. *1962*, 214.
14. a) Clauss, K.: Liebigs Ann. Chem. *711*, 19 (1968); b) Rausch, M. D., Gordon, H. B.: J. Organomet. Chem. *74*, 85 (1974).
15. Steinbach, R.: Dissertation, Univ. Marburg 1983.
16. Besides lit. [10], additional recent reviews of titanium chemistry have appeared: a) Reetz, M. T.: Nachr. Chem. Techn. Lab. *29*, 165 (1981); b) Weidmann, B., Seebach, D.: Angew. Chem. *95*, 12 (1983); Angew. Chem., Int. Ed. Engl. *22*, 31 (1983); c) Bottrill, M., Gavens, P. D., Kelland, J. W., McMeeking, J.: in "Comprehensive Organometallic Chemistry", Wilkinson, G., Stone, F. G. A., Abel, E. W. (editors), Pergamon Press, Oxford, Chapter 22, 1982; d) Reetz, M. T.: Pure Appl. Chem. *57*, 1781 (1985); e) see also lit. [31b]
17. Ostarek, R.: Diplomarbeit, Univ. Marburg 1983.
18. Weidmann, B., Widler, L., Olivero, A. G., Maycock, C. D., Seebach, D.: Helv. Acta *64*, 357 (1981).
19. Wenderoth, B.: Dissertation, Univ. Marburg 1983.
20. Kauffmann, T., Antfang, E., Ennen, B., Klas, N.: Tetrahedron Lett. *23*, 2301 (1982).
21. Reetz, M. T., Steinbach, R., Kesseler, K.: Angew. Chem. *94*, 872 (1982); Angew. Chem., Int. Ed. Engl. *21*, 864 (1982); Angew. Chem. Supplement *1982*, 1899.
22. Reetz, M. T., et al.: unpublished results 1982–85.
23. Reetz, M. T., Kyung, S. H., Hüllmann, M.: Tetrahedron, in press.
24. Compound *64* is also accessible by treating a solution of *17* (made from $(CH_3)_2Zn$ and $TiCl_4$ in CH_2Cl_2) with ether. The analogous bis-THF complex can be made similarly. These octahedral complexes react chemo- and stereoselectively: Reetz, M. T., Westermann, J.: Synth. Commun. *11*, 647 (1981).
25. Kauffmann, T., König, R., Pahde, C., Tannert, A.: Tetrahedron Lett. *22*, 5031 (1981).

26. a) Review of the Peterson olefination: Ager, D. J.: Synthesis *1984*, 384; see also b) Cohen, T., Sherbine, J. P., Matz, J. R., Hutchins, R. R., McHenry, B. M., Willey, P. R.: J. Am. Chem. Soc. *106*, 3245 (1984).

27. a) Improved synthesis of ClTi(NEt$_2$)$_3$ and Ti(NEt$_2$)$_4$: Reetz, M. T., Urz, R., Schuster, T.: Synthesis *1983*, 540; b) review of aminotitanium compounds: Bürger, H., Neese, H. J.: Chimia *24*, 209 (1970).

28. a) Bürger, H., Neese, H. J.: J. Organomet. Chem. *20*, 129 (1969); b) Bürger, H., Neese, H. J.: J. Organomet. Chem. *21*, 381 (1970); c) Neese, H. J., Bürger, H.: J. Organomet. Chem. *32*, 213 (1971); d) Bürger, H., Neese, H. J.: J. Organomet. Chem. *36*, 101 (1972).

29. Reetz, M. T., Wenderoth, B.: Tetrahedron Lett. *23*, 5259 (1982).

30. Seebach, D., Schiess, M.: Helv. Chim. Acta *65*, 2598 (1982).

31. a) Seebach, D., Beck, A. K., Schiess, M., Widler, L., Wonnacott, A.: Pure Appl. Chem. *55*, 1807 (1983); b) Seebach, D. in: "Modern Synthetic Methods", (Scheffold, R., editor), Vol. III, Salle Verlag, Frankfurt; Verlag Sauerländer, Aarau, p. 217, 1983.

32. Reetz, M. T., Peter, R.: Tetrahedron Lett. *22*, 4691 (1981).

33. Hanko, R., Hoppe, D.: Angew. Chem. *94*, 378 (1982); Angew. Chem., Int. Ed. Engl. *21*, 372 (1982); Angew. Chem. Supplement *1982*, 961.

34. Reetz, M. T., Kyung, S. H., Westermann, J.: Organometallics *3*, 1716 (1984).

35. a) Kükenhöhner, T.: Diplomarbeit, Univ. Marburg 1983; b) Erskine, G. J., Hunter, B. K., McCowan, J. D.: Tetrahedron Lett. *26*, 1371 (1985).

36. Reetz, M. T., Steinbach, R., Westermann, J., Peter, R., Wenderoth, B.: Chem. Ber. *118*, 1441 (1985).

37. Reetz, M. T., Wenderoth, B., Peter, R.: J. Chem. Soc., Chem. Commun. *1983*, 406.

38. Kauffmann, T., Möller, T., Rennefeld, H., Welke, S., Wieschollek, R.: Angew. Chem. *97*, 351 (1985); Angew. Chem., Int. Ed. Engl. *24*, 348 (1985).

39. a) Schöllkopf, U.: in Houben-Weyl-Müller, "Methoden der Organischen Chemie", Vol. 13/1, p. 87, Thieme-Verlag, Stuttgart 1970: b) Stowell, J. C.: "Carbanions in Organic Synthesis", Wiley, N.Y. 1979); c) Posner, G. H.: Org. React. *19*, 1 (1972) and *22*, 253 (1975); d) Lipshutz, B. H., Wilhelm, R. S., Kozlowski, J. A.: Tetrahedron *40*, 5005 (1984).

40. Reviews of these and other aspects of allylsilane reagents: a) Sakurai, H.: Pure Appl. Chem. *54*, 1 (1982); b) Sakurai, H., Hosomi, A., Hayashi, J.: Org. Synth. *62*, 86 (1984); c) Chan, T. H., Fleming, I.: Synthesis *1979*, 761; d) Parnes, Z. N., Bolestova, G. I.: Synthesis *1984*, 991; e) W. P. Weber: "Silicon Reagents for Organic Synthesis", Springer-Verlag, Berlin 1983.

41. Roulet, D., Casperos, J., Jacot-Guillarmod, A.: Helv. Chim. Acta *67*, 1475 (1984).

42. Similar phenomena have been observed in certain hydride reductions: Noyori, R.: private communication.

43. Brun, L.: Acta Crystallogr. *20*, 739 (1966).

44. Reetz, M. T., Wenderoth, B., Urz, R.: Chem. Ber. *118*, 348 (1985).

45. a) Dornow, A., Gehrt, H., Ische, F.: Liebigs Ann. Chem. *585*, 220 (1954); b) Brière, R., Rassat, A.: Bull. Soc. Chim. Fr. *1965*, 378; c) Bartoli, G., Bosco, M., Dalpozzo, R.: Tetrahedron Lett. *26*, 115 (1985).

46. Kostova, K., Hesse, M.: Helv. Chim. Acta *67*, 1713 (1984).

47. In the paper by Hesse [46], intermediates *169–171* are formulated with ionic Ti—O bonds.

48. Kostova, K., Lorenzi-Riatsch, A., Nakashita, Y., Hesse, M.: Helv. Chim. Acta *65*, 249 (1982).

49. Kostova, K., Hesse, M.: Helv. Chim. Acta *66*, 741 (1983).

50. Aono, T., Hesse, M.: Helv. Chim. Acta *67*, 1448 (1984).

51. $TiCl_4/CH_2=CHCH_2SiMe_3$ is an excellent reagent combination for chelation-controlled additions to chiral alkoxy aldehydes; review: Reetz, M. T.: Angew. Chem. *96*, 542 (1984); Angew. Chem., Int. Ed. Engl. *23*, 556 (1984).

52. Reetz, M. T., Steinbach, R., Westermann, J., Urz, R., Wenderoth, B., Peter, R.: Angew. Chem. *94*, 133 (1982); Angew. Chem., Int. Ed. Engl. *21*, 135 (1982); Angew. Chem. Supplement *1982*, 257.

53. Jung, A.: Diplomarbeit, Univ. Marburg 1983.

54. Kauffmann, T., Abeln, R., Wingbehrmühle, D.: Angew. Chem. *96*, 724 (1984); Angew. Chem., Int. Ed. Engl. *23*, 729 (1984).

55. Nützel, K.: in Houben-Weyl-Müller, "Methoden der Organischen Chemie", Vol. 13/2a, p. 47, Thieme-Verlag, Stuttgart 1973.

56. Weidmann, B., Maycock, C. D., Seebach, D.: Helv. Chim. Acta *64*, 1552 (1981).

57. Maier, G.: private communication to M. T. Reetz, 1984.

58. a) Reetz, M. T., Weis, C.: Synthesis *1977*, 135; b) Reetz, M. T., Stephan, W.: J. Chem. Res. (S), *1981*, 44; J. Chem. Res. (M), *1981*, 583; c) Reetz, M. T., Schinzer, D.: Angew. Chem. *89*, 46 (1977); Angew. Chem., Int. Ed. Engl. *16*, 44 (1977); d) Reetz, M. T., Stephan, W.: Liebigs Ann. Chem. *1980*, 171.

59. Cardin, D. J., Norton, R. J.: J. Chem. Soc., Chem. Commun. *1979*, 513.

60. Tabusa, F., Yamada, T., Suzuki, K., Mukaiyama, T.: Chem. Lett. *1984*, 405.

61. Prof. M. Braun (Düsseldorf) has made similar observations: private communication to M. T. Reetz, 1984.

62. Molecular orbital calculations: Wolfe, S., LaJohn, L. A., Weaver, D. F.: Tetrahedron Lett. *25*, 2863 (1984).

63. a) Nützel, K.: in Houben-Weyl-Müller, "Methoden der Organischen Chemie", Vol. 13/2a, p. 553, Thieme-Verlag, Stuttgart 1973; b) Nützel, K.: in Houben-Weyl-Müller, "Methoden der Organischen Chemie", Vol. 13/2a, p. 859, Thieme-Verlag, Stuttgart 1973.

64. Cahiez, G., Bernard, D., Normant, J. F.: Synthesis *1977*, 130.

65. Cahiez, G., Normant, J. F.: Tetrahedron Lett. *18*, 3383 (1977).

66. Posner, G. H., Whitten, C. E., McFarland, P. E.: J. Am. Chem. Soc. *94*, 5106 (1972).

67. Yamamoto, Y., Yamamoto, S., Yatagai, H., Ishihara, Y., Maruyama, K.: J. Org. Chem. *47*, 119 (1982).

68. Naruta, Y., Ushida, S., Maruyama, K.: Chem. Lett. *1979*, 919.

69. a) Hoffmann, R. W.: Angew. Chem. *94*, 569 (1982); Angew. Chem., Int. Ed. Engl. *21*, 555 (1982); b) Yamamoto, Y., Maruyama, K.: Heterocycles *18*, 357 (1982).

70. Okude, Y., Hirano, S., Hiyama, T., Nozaki, H.: J. Am. Chem. Soc. *99*, 3179 (1977).

71. Kauffmann, T., Hamsen, A., Beirich, C.: Angew. Chem. *94*, 145 (1982); Angew. Chem., Int. Ed. Engl. *21*, 144 (1982).

72. Buse, C. T., Heathcock, C. H.: Tetrahedron Lett. *19*, 1685 (1978).

73. Imamoto, T., Sugiura, Y., Takiyama, N.: Tetrahedron Lett. *25*, 4233 (1984).

74. See for example, a) Negishi, E.: "Organometallics in Organic Synthesis", Wiley, N.Y. 1980; b) Davies, S. G.: "Organotransition Metal Chemistry: Applications to Organic Synthesis", Pergamon Press, Oxford 1982; c) Collman, J. P., Hegedus, L. S.: "Principles and Applications of Organotransition Metal Chemistry", University Science Books, Mill Valley, California 1980.

75. a) Girard, P., Namy, J. L., Kagan, H. B.: J. Am. Chem. Soc. *102*, 2693 (1980);
    b) Namy, J. L., Souppe, J., Kagan, H. B.: Tetrahedron Lett. *25*, 765 (1984);
    c) Ananthanaryan, T. P., Gallagher, T., Magnus, P.: Tetrahedron Lett. *23*,
    3497 (1982); d) Natale, N. R.: Tetrahedron Lett. *23*, 5009 (1982).
76. Weingarten, H., White, W. A.: J. Org. Chem. *31*, 4041 (1966).
77. Hüllmann, M.: Diplomarbeit, Univ. Marburg 1983.
78. Greene, T. W.: "Protective Groups in Organic Synthesis", Wiley, N.Y. 1981.
79. a) Comins, D. L., Brown, J. D.: Tetrahedron Lett. *22*, 4213 (1981); b) Comins,
    D. L., Brown, J. D., Mantlo, N. B.: Tetrahedron Lett. *23*, 3979 (1982).
80. Vuitel, L., Jacot-Guillarmod, A.: Synthesis *1972*, 608.
81. Crociani, B., Nicolini, M., Richards, R. L.: J. Organomet. Chem. *101*, C 1 (1975).
82. Schiess, M., Seebach, D.: Helv. Chim. Acta *66*, 1618 (1983).
83. a) Passerini, M.: Gazz. Chim. Ital. *51*, II, 181 (1921); b) Ugi, I.: Angew. Chem.
    94, 826 (1982); Angew. Chem., Int. Ed. Engl. *21*, 810 (1982).
84. Hagedorn, I., Eholzer, U.: Chem. Ber. *98*, 936 (1965).
85. a) Nakamura, E., Kuwajima, I.: J. Am. Chem. Soc. *105*, 651 (1983); b) Naka-
    mura, E., Kuwajima, I.: J. Am. Chem. Soc. *107*, 2138 (1985).
86. Reissig, H. U.: Tetrahedron Lett. *22*, 298 (1981).

4. Rates of Reactions

Inspite of the fact that many organotitanium compounds are well characterized, the number of kinetic studies is limited. This chapter deals primarily
with such efforts directed toward elucidating the intricacies of carbonyl
addition of $CH_3Ti(OCHMe_2)_3$. Various other kinetic processes are also
briefly discussed (Section 4.2).

4.1 Kinetics of the Addition of $CH_3Ti(OCHMe_2)_3$ to Carbonyl Compounds

Since the Grignard reaction is of fundamental synthetic importance, much
research has centered around the mechanism of carbonyl addition. The research groups of Smith [1], Ashby [2], Holm [3] and others have reported a great
deal of mechanistic data involving RMgX and RLi, including those of kinetic
experiments. The complexities of carbonyl addition have been partially
unraveled. Thus, the effects of the Schlenk equilibrium, aggregation, solvation, electron transfer processes, etc. are now fairly well understood.

Progress in titanium chemistry has not reached this level, inspite of the
fact that the experimental problems are not as pronounced. For example,
$CH_3Ti(OCHMe_2)_3$ is a distillable reagent (Chapter 2) which is easily handled.
Furthermore, it is much less reactive than CH_3Li or CH_3MgX, so that reliable
kinetic data can be obtained using conventional techniques. Nevertheless, the
compound does tend to form aggregates (e.g., dimers) via Ti—O bridging to
some extent, depending upon concentration and temperature [4] (Chapter 2).
Such measurements were performed cryoscopically in benzene at about
$+5\,°C$. Since addition of $CH_3Ti(OCHMe_2)_3$ to aldehydes under such conditions is too fast to be monitored kinetically using conventional methods,
lower temperatures and different solvents, e.g., (CH_2Cl_2 and THF) have to be
employed. Thus, care must be taken in comparing the various systems. Whereas molecular weight determinations have not yet been carried out in CH_2Cl_2 or
THF at low temperatures, ^{13}C-NMR studies of $CH_3Ti(OCHMe_2)_3$ in CH_2Cl_2
in the temperature range $-50\,°C$ to $+30\,°C$ clearly show the existence of
several aggregated species (Chapter 2) [5].

The above factors should be kept in mind when considering the kinetic
data which is currently available [5]. The rates of addition of
$CH_3Ti(OCHMe_2)_3$ to heptanal at various temperatures in the range $-20\,°C$

to $-65\,°C$ were measured at $5°$ intervals in CH_2Cl_2 and THF. Typically, 0.3 M solutions of organotitanium reagent were used. The addition is extremely clean, conversion to *4* being $>95\%$. Samples of the reaction mixture were periodically hydrolyzed and analyzed by capillary gas chromatography.

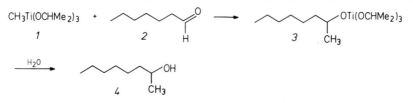

In all cases good adherence to second-order rate law was observed (first-order with respect to reagent and to substrate). The activation parameters turned out to be as follows [5]:

Solvent	$\Delta G^{\ddagger}$ (kJ/mol)	$\Delta H^{\ddagger}$ (kJ/mol)	$\Delta S^{\ddagger}$ $(J \cdot K^{-1} \cdot mol^{-1})$
CH_2Cl_2	70.8 ± 5.4	51.9 ± 5.0	-80.8 ± 16.7
THF	69.1 ± 5.4	85.8 ± 5.0	$+72.0 \pm 16.7$

The results show that the rate of addition is essentially solvent-independent, but that THF participation is nevertheless involved. Whereas $\Delta S^{\ddagger}$ in case of CH_2Cl_2 is negative as expected for a bimolecular reaction, it is positive when THF is used as the solvent. The effect is compensated by the different $\Delta H^{\ddagger}$ values, leading to almost identical activation energies.

Although a final discussion concerning these numbers must await further experimentation, the following hypothesis is in line with the data. Solvated reagent as shown in *5* (two THF molecules arbitrarily assumed) must first kick off the THF to form free *1* before reacting with n-heptanal (*2*). This is reflected in the high $\Delta H^{\ddagger}$ value as well as in the positive $\Delta S^{\ddagger}$. Related solvent effects have been observed in other cases [6], most recently in the Ivanov reaction [7]. The simplified scheme neglects dimeric or aggregated forms of the titanium reagent, which are in rapid equilibrium with the monomeric form. Also, the equilibrium $5 \leftrightarrows 1$ lies to the left only in case of a large excess of THF (i.e., as solvent). The use of two equivalents of THF per equivalent of reagent in case of the addition in CH_2Cl_2 has no effect on the activation parameters [5].

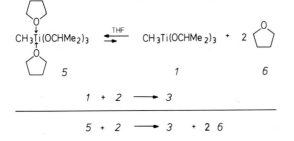

Whatever the final interpretation, the results show that $CH_3Ti(OCHMe_2)_3$ is much less reactive than CH_3Mg or CH_3Li. From a synthetic point of view, the important conclusion is that the choice of solvent is not important. Initial kinetic studies using non-protic solvents having different polarity parameters E_t [8] substantiate this [5] (Table 1).

Table 1. Approximate Conversion in the Reaction $1 + 2 \rightarrow 4$ at $-41\ °C$ after 1 Hour

Solvent	E_t-Value	Conversion $(\%)^a$
CH_2Cl_2	44.1	25
THF	37.4	30
Ether	34.6	32
Toluene	33.9	26
n-Hexane	30.9	33

[a] As measured by the formation of *4* following aqueous workup.

Preliminary kinetic studies concerning the addition of the tri-ethoxy derivative *7* to *2* reveal a more complicated situation [5]. In CH_2Cl_2 the reagent is less reactive than *1* by a factor of about 40 (at $-30\ °C$). Also, at temperatures below $-10\ °C$, no adherence to the usual second-order rate law was observed. Thus, the dimeric or more highly aggregated (at low temperatures) forms of *7*, which are considerably more stable than those of *1* (see Chapter 2 concerning aggregation), are responsible for the low reactivity of this system [9]. Interestingly, second-order kinetics were observed in case of THF as the solvent [5].

$$CH_3Ti(OEt)_3 \qquad CH_3Ti(O\text{-}t\text{-}Bu)_3$$
$$7 \qquad\qquad\qquad 8$$

Thus, in choosing alkoxy ligands at titanium, it is best to consider groups which are at least as large as isopropoxy. It remains to be seen how the kinetics of the bulky and monomeric *t*-butoxy derivative *8* turn out.

Currently, it is not clear how $CH_3Ti(OCHMe_2)_3$ adds to an aldehyde, directly via a four-center transition state *9*, or in a two step sequence in which complexation according to *11* precedes C—C bond formation. In case of RLi and RMgX, spectroscopic evidence for such complexation has been presented [1–3, 10]. So far, this information is not available for titanium reagents. RO-adducts in equilibrium with the free aldehyde should also be considered in reactions of $CH_3Ti(OR)_3$.

Irrespective of such details, the allyl analog *12* reacts much faster than *1*, as judged by qualitative observations. Although detailed kinetic studies need to be carried out, the effect upon going from *1* to *12* is so pronounced that something fundamentally different must be occurring. Indeed, crotyl-

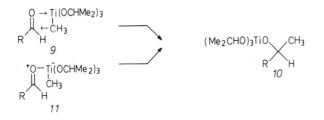

titanium reagents react with allyl inversion, which is best explained by a six-membered cyclic transition state (Chapter 5). This may well be stereo-electronically more favorable. If complexation occurs at the lone-electron pair of oxygen, the end of the allyl group can "reach" the *p*-orbital of the carbon atom much easier than a methyl group. This may apply to other allyl- vs. methylmetal reagents as well. It would be interesting to measure the activation parameters in such cases. The possible role of orbital symmetry also needs to be clarified. *n*-Alkyltitanium reagents are slightly less reactive than the parent compound *1*, whereas "resonance-stabilized" species such as titanated enolates *13*, nitriles *14*, sulfones *15* etc. react faster [9]. It should be noted that the structure of *14* and *15* is not known, i.e., titanium may be attached to nitrogen and oxygen, respectively (see Chapter 2).

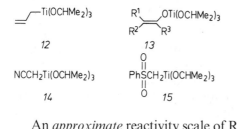

An *approximate* reactivity scale of RTiX$_3$ (X = OCHMe$_2$) based on kinetic and in part qualitative experiments is as follows [9]:

$$CH_2=CHCH_2TiX_3 \geq NC-CH_2TiX_3 \gg CH_3TiX_3 \sim PhTiX_3$$
$$> n\text{-}C_4H_9TiX_3 > Me_3SiCH_3TiX_3$$

The list is incomplete, but does provide a guide in synthetic applications. For example, titanium reagents having two different carbon moieties react selectively (e.g., *16* → *17*) [11].

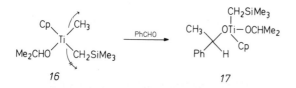

Upon substituting isopropoxy ligands in *1* for methyl groups, reactivity toward carbonyl compounds increases dramatically [9, 12]. For example, the addition of the parent compound *1* to ketones (e.g., 2-hexanone) requires

room temperature and fairly long reaction times ($\sim$15 h), whereas the analogous process with *18* occurs smoothly at 0 °C within three hours. In the
latter reaction a 1:1 ratio of components is used so that only one active
methyl group is utilized. The intermediate *20* can either be quenched with
H$_2$O/H$^+$, or be used in a second addition (which is slower). Increase in
reactivity is even more drastic upon going to (CH$_3$)$_4$Ti [13]. This is easily
understood by considering the strong Ti—O bond as well as steric factors
(Chapter 2).

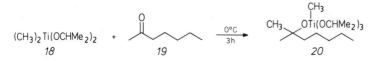

A similar increase in the rate of carbonyl addition occurs when the
alkoxy ligands are replaced by chlorine groups. For example, CH$_3$TiCl$_3$
(generated by CH$_3$Li/TiCl$_4$ in ether; see Chapter 3) adds smoothly to ketones
in the temperature range -20 °C to 0 °C [14]. Since *21* (in this medium in
equilibrium with etherates; see Chapter 2) is also very chemoselective, it is
often the reagent of choice relative to *1* or *18*.

CH$_3$TiCl$_3$ $\quad\xrightarrow[\text{2. H}_2\text{O}]{\text{1. }19 \text{ /}-10°\text{C / 3h}}\quad$

21   *22*

Pentahapto cyclopentadienyl groups are strongly electron releasing ligands (Chapter 2) which slow down the carbonyl addition reactions [9, 15].
For example, *23* reacts sluggishly with aldehydes, and *24* fails to add at all
at room temperature (48 h) [16]. The non-cyclic mono-Cp derivatives *25* and
26 are more reactive, however, adding to aldehydes at 0 °C to +22 °C
[16–19]. *26* also reacts with ketones, e.g., stereoselectively with 4-*tert*-butyl-
cyclohexanone (Chapter 5). In all of these systems, more reactive carbon nucleophiles such as allyl or enolate moieties can in fact be used synthetically
[9, 15, 16].

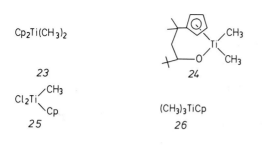

Cp$_2$Ti(CH$_3$)$_2$

23

Cl$_2$Ti(CH$_3$)(Cp)

25

(CH$_3$)$_3$TiCp

26

A synthetically important aspect of titanating classical carbanions concerns the increase in chemoselectivity (Chapter 3). For example, reagents
RTi(OCHMe$_2$)$_3$ and RTiCl$_3$ are aldehyde-selective in the presence of
ketones, or ketone-selective in the presence of esters and other functionalities.

The competition experiments described in Chapter 3 do not lead to precise kinetic data. For example, a 1:1:1 mixture of $CH_3Ti(OCHMe_2)_3$, an aldehyde and a ketone affords only the aldehyde adduct as shown by GC, but this only gives a lower limit of the relative rate [9]. A rate factor of 100–150 is all that synthetic organic chemists need; nevertheless, it was of interest to determine the precise number.

Using the parent compound $CH_3Ti(OCHMe_2)_3$ and various aldehyde/aldehyde, aldehyde/ketone and ketone/ketone pairs, the relative rates of carbonyl addition for aldehyde/ketone pairs were determined at room temperature [5, 9]. Depending upon the particular aldehyde/ketone pair chosen, the k_{rel} values varied between 220 and 700:

$$k_{rel} = k_{aldehyde}/k_{ketone} = 220-700$$

For example, $k_{heptanal}/k_{3-heptanone} = 223$

and $k_{benzaldehyde}/k_{acetophenone} = 550$.

At low temperatures these numbers are likely to be considerably higher [5]. Obviously, if very bulky aldehydes are used, k_{rel} may turn out to be less than 200. Even aldehyde/aldehyde and ketone/ketone discrimination may amount to k_{rel} values of 10 to 40 (Chapter 3).

These and other results (Chapter 3) demonstrate that titanium reagents are very sensitive to changes in steric and electronic properties of the carbonyl compounds. Initial Hammett-type studies reveal that electron-withdrawing substituents increase the rate of carbonyl addition [5]:

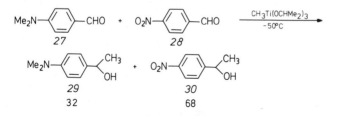

4.2 Other Kinetic Studies

Several other kinetic investigations involving reactions other than carbonyl additions have been reported [20]. A few of them are mentioned here briefly. For example, the $\Delta G^{\ddagger}$ of enantiomerization of *31* amounts to 19.2 kcal/mol (80.3 kJ/mol) [15], as determined by dynamic ^{1}H-NMR spectroscopy (coalescence of the diastereotopic methyl groups). The mechanism is unclear, but may involve intermediate Ti—O—Ti bridging. Redistribution to form new species does not occur under the reaction conditions.

CH₃—Ti with OCHMe₂, Cp, Cl
31

The rates of redistribution reactions involving ligand exchange processes at titanium have been reported [21, 22]:

$$TiX_4 + TiY_4 \rightleftarrows TiX_3Y + TiX_2Y_2 + TiXY_3$$

$$32 \qquad 33 \qquad 34 \qquad 35 \qquad 36$$

$$X, Y = Cl, NR_2, OR$$

For example, dimethylamino and *t*-butoxy groups were scrambled in a second-order reaction in toluene (first-order in both *37* and *38*). The rate was ascertained by measuring the appearance of the first scrambling product *39*, for which the rate constant turned out to be $k = (4.2 \pm 0.4) \cdot 10^{-5}$ l $\times mol^{-1} sec^{-1}$. The exchange process has an activation enthalpy of 9.6 kcal/mol (40 kJ/mol) and an activation entropy of -46.9 cal $\cdot K^{-1} mol^{-1}$ (196 JK^{-1} mol^{-1}). The latter is consistent with a four-center transition state [21].

$$Ti(NMe)_4 + Ti(O\text{-}t\text{-}Bu)_4 \leftrightarrows t\text{-}BuOTi(NMe_2)_3 + Me_2NTi(O\text{-}t\text{-}Bu)_3$$

$$37 \qquad\qquad 38 \qquad\qquad 39 \qquad\qquad\qquad 40$$

Interestingly, the rate of scrambling is five orders of magnitude greater if the less bulky isopropoxy analog of *38* is used [21]. The preexchange lifetimes of the species in *32* + *33* ⇆ *34* + *35* + *36* in case of chlorides in combination with alkoxides and amides are also very short [22]. The reader is referred to the original literature for rate data, equilibrium values as well as heats of redistribution [21–23]. Exchange is slower in case of Cp-derivatives [24], which reflects the electron-donating and steric properties of such ligands (see Chapter 2). These phenomena must be kept in mind when attempting to synthesize compounds having a center of chirality at titanium (Chapter 5). It should be mentioned that often not all possible species are actually formed, i.e., redistribution may rapidly lead to a single product (Chapter 2), e.g., *42*:

$$2\ CH_3Ti(OR)_3 + CH_3TiCl_3 \rightleftharpoons 3\ CH_3Ti(OR)_2Cl$$

$$41 \qquad\qquad 21 \qquad\qquad 42$$

The rate of methyl exchange between CH_3TiCl_3 and $(CH_3)_2Zn$ in C_2Cl_4 and benzene has been measured by NMR spectroscopy ($\Delta G^{\neq} = 7.5$ kcal/mol (31.4 kJ/mol)) [25]. Methyl exchange between CD_3TiCl_3 and $(CH_3)_2TiCl_2$ is slow on the NMR time scale at room temperature, but fast at $+115\ °C$ [25].

The kinetics of decomposition of several organotitanium reagents have been studied (see also Chapter 2). For example, decay of CH_3TiCl_3 in the presence of $Al(C_2H_5)_3$ to form $TiCl_3$ and CH_4 has an activation energy of 11 kcal/mol (46 kJ/mol) [26]. Decomposition of CH_3TiCl_3 in ether follows a second-order rate law and leads to $TiCl_3(OEt_2)_2$ (*44*): the rate constant amounts to $k = 0.43 \cdot 1 \cdot mol^{-1} \, min^{-1}$ (at $+273 \, °K$) and $1.36 \cdot 1 \cdot mol^{-1}$ $\times min^{-1}$ (at 298 °K) [27].

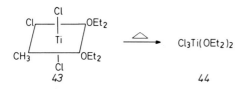

$$Cl_3Ti(OEt_2)_2$$

44

The first step of the decomposition of *45* at room temperature has an activation energy of $15 \pm 3 \, kcal/mol$ $(62 \pm 12 \, kJ/mol)$ [28]. It should be mentioned that decomposition of certain organotitanium compounds can be affected by impurities or autocatalysis, so that reliable kinetic data may not be easily accessible in some cases. More work is needed in the area of decomposition.

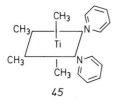

45

Turning to completely other chemical processes, the kinetics of a degenerate titanium mediated olefin metathesis have been determined [29]. Dynamic processes in tetra-cyclopentadienyltitanium (*46*) [30] and bis (β-diketonato)-titanium(IV) compounds *47* [31] have been studied using NMR spectroscopy. Complex *47* exist as rapidly interconverting diastereomers; the activation parameters are in line with a twist mechanism as opposed to a process involving dissociation/association [31].

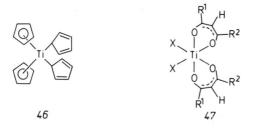

46   47

Finally, an important study concerning cleavage reactions of methylaryltitanium(IV) compounds *48* induced by electrophiles of the type HCl, HOAc, $HgCl_2$ and CH_3HgCl has appeared [32]. Both HCl and HOAc show a slight preference for phenyl rather than methyl cleavage, while the mercury compounds displays opposite selectivity. Interesting substituent effects were also observed, e.g.:

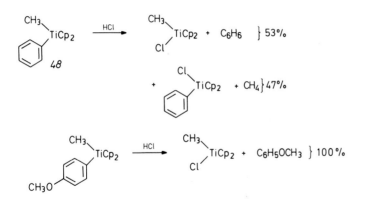

Based on these and other experiments, including rate studies (Hammett correlations), it was concluded that the mechanism not always involves S_E2 cleavage. For example, in case of $HgCl_2$, an electron transfer mechanism appears to be operating [32].

References

1. a) Smith, S. G., Charbonneau, L. F., Novak, D. P., Brown, T. L.: J. Am. Chem. Soc. 94, 7059 (1972); b) Al-Aseer, M. A., Smith, S. G.: J. Org. Chem. 49, 2608 (1984), and lit. cited therein.
2. a) Ashby, E. C., Laemmle, S., Neumann, H. M.: Acc. Chem. Res. 7, 272 (1974); b) Ashby, E. C.: Pure Appl. Chem. 52, 545 (1980).
3. a) Holm, T.: Acta Chem. Scand. 23, 1829 (1969); b) Holm, T.: Acta Chem. Scand. 25, 833 (1971); c) Holm, T.: Acta Chem. Scand. B 37, 567 (1983).
4. Kühlein, K., Clauss, K.: Makromol. Chem. 155, 145 (1972).
5. a) Maus, S.: Diplomarbeit, Univ. Marburg 1983; b) Maus, S.: projected Dissertation, Univ. Marburg 1986.
6. Reichardt, C.: "Solvent Effects in Organic Chemistry", Verlag Chemie, Weinheim 1979.
7. Toullec, J., Mladenova, M., Gaudemar-Bardone, F., Blagoev, B.: Tetrahedron Lett. 24, 589 (1983).
8. Reichardt, C.: Liebigs Ann. Chem. 752, 64 (1971).
9. Reetz, M. T.: Top. Curr. Chem. 106, 1 (1982).
10. Lozach, D., Molle, H., Bauer, P., Dubois, J. E.: Tetrahedron Lett. 24, 4213 (1984).
11. Westermann, J.: Dissertation, Univ. Marburg 1982.
12. Reetz, M. T., Westermann, J., Steinbach, R., Wenderoth, B., Peter, R., Ostarek, R., Maus, S.: Chem. Ber. 118, 1421 (1985).
13. Reetz, M. T., Steinbach, R., Westermann, J., Urz, R., Wenderoth, B., Peter, R.: Angew. Chem. 94, 133 (1982); Angew. Chem., Int. Ed. Engl. 21, 135 (1982); Angew. Chem. Supplement 1982, 257.
14. Reetz, M. T., Kyung, S. H., Hüllmann, M.: Tetrahedron, in press.
15. Reetz, M. T., Kyung, S. H., Westermann, J.: Organometallics 3, 1716 (1984).
16. Kükenhöhner, T.: Diplomarbeit, Univ. Marburg 1983.
17. Steinbach, R.: Dissertation, Univ. Marburg 1982.

acidic titanium compounds. For example, ¹H-NMR studies of *2* show that upon changing the solvent from tetrachloroethylene to benzene, the methyl signal shifts upfield from $\delta = 2.47$ to 2.00 [5], in line with benzene-titanium complexation (Chapter 2). Furthermore, it was known that TiCl₄ forms charge-transfer complexes with aromatic compounds [6] (analogous investigations using *2* remain to be carried out). The fact that such compounds as *2*, TiCl₄ or CH₃TiCl₃ (*9*) form six-coordinate octahedral complexes with THF (e.g., *10*), ether, amines and related bidentate ligands is also noteworthy (Chapter 2). Finally, the X-ray crystallographic structure of tetrabenzyltitanium (which is also a Lewis acid) shows strong intramolecular interaction between titanium and the phenyl π-face (Chapter 2).

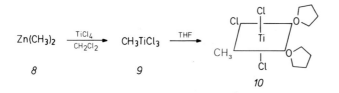

$$
\text{Zn(CH}_3)_2 \xrightarrow[\text{CH}_2\text{Cl}_2]{\text{TiCl}_4} \text{CH}_3\text{TiCl}_3 \xrightarrow{\text{THF}}
$$

8   9   10

Many of the above phenomena do not apply to titanium reagents having ligands other than chlorine. For example, CH₃Ti(OCHMe₂)₃ (*12*) does not form a stable *bis*-THF adduct analogous to *10*, because the π-donor property of the alkoxy ligands reduces Lewis acidity drastically (Chapter 2). *12* is also fairly bulky. It does not react with *1* in the same way as *2*.

$$
\text{ClTi(OCHMe}_2)_3 \xrightarrow{\text{CH}_3\text{Li}} \text{CH}_3\text{Ti(OCHMe}_2)_3
$$

11   12

The stereoselective addition of *2* to *1* set the stage for systematically testing organotitanium reagents in stereoselective reactions. It seemed that the type of the ligand at titanium could determine the electronic and steric nature of the reagent [8] (Chapter 1). These expectations were later fulfilled. Indeed, an instrument is now at hand which allows the generation of the tailor-made reagents from classical organometallics such as RLi, RMgX, RZnX, ZnR₂ as well as a host of traditional "carbanions". For example, CH₃TiCl₃ (*9*) should how efficient chelation-control in addition reactions of chiral alkoxy aldehydes, while CH₃Ti(OCHMe₂)₃ (*12*) might be expected to form non-chelation-controlled products.

It turned out that in case of other stereochemical problems, e.g., Cram/anti-Cram selectivity or axial/equatorial addition to cyclic ketones, the optimum ligand system is not always easily predicted. For this reason empirical rules had to be established [8]. The bulk of the ligands (e.g., various alkoxy or amino groups) often influences stereoselectivity in a predictable way. The following sections are devoted to the influence of titanation on stereoselectivity. Included are also TiCl₄ mediated stereoselective C—C bond forming reactions.

5.2 Diastereofacial Selectivity

The two π-faces of an aldehyde or ketone with at least one chiral center are diastereotopic. Thus, addition of C-nucleophiles such as RMgX, RLi or enolates can lead to unequal amounts of diastereomers. Reactions involving such 1,n-asymmetric induction [4] have been termed diastereofacially selective [1]. Although this phenomenon was first observed some 90 years ago [4], it was not until the pioneering work of Cram that a certain degree of systematization was attempted [4, 9]. In what is now known as Cram's rule, an α-chiral aldehyde (or ketone) such as *13* is assumed to adobt a conformation in which the largest of the three α-substituents is antiperiplanar to the carbonyl function, nucleophilic attack then occurring preferentially from the less hindered side. *14* is the so-called Cram-product, *15* the anti-Cram product. *13* is arbitrarily shown in one enantiomeric form, although racemates are usually used, leading to racemic products *14* and *15*.

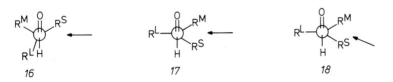

R^S = small, R^M = medium, R^L = large substituents

Later, the Cram model (cf. Newman projection *16*) was refined by Felkin (cf. *17*) [10] and subsequently by Anh (cf. *18*) [11].

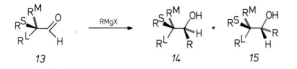

For reactions of chiral carbonyl compounds having α-halogen substituents, Cornforth proposed an electrostatic model in which the electronegative halogen points away from the polar carbonyl function, i.e., halogen takes the place of R^L in *16* [12]. In contrast, Felkin postulated that polar effects stabilize those transition states in which the separation between the incoming nucleophile and the electronegative α-substituent is greatest [10], as in *17* (R^L = electronegative group). Anh's MO calculations of α-chloropropanal show that such a conformation is in fact the most reactive form of the molecule because $\pi^*_{C=O} - \sigma^*_{C-Cl}$ interaction provides a low-lying LUMO; attack anti to the electronegative substituent (e.g., Cl) at an angle $>90\%$ according to *18* (R^L = electronegative substituent) is then energetically most favorable [11].

In case of α-alkoxy or hydroxy carbonyl compounds the electronegative oxygen can potentially exert analogous effects. However, a different phenomenon is also possible, namely chelation, which makes the opposite dia-

5. Stereoselectivity in the Addition of Organotitanium Reagents

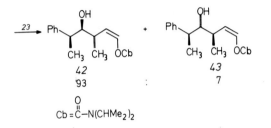

$$Cb = \overset{O}{\underset{\|}{C}} - N(CHMe_2)_2$$

   The aldehyde *23* has been reacted with a number of enolates, with varying degrees of success [14]. An excellent method to obtain Cram products in the aldol reaction involves the BF_3-promoted addition of enol silanes; for example, reactions of *23* lead to Cram/anti-Cram product ratios of >15: <1 [16]. Titanium enolates [26] have not yet been tested with α-chiral aldehydes devoid of heteroatoms.
   Besides *23*, other chiral aldehydes also react stereoselectively with titanium reagents. 2-Phenylbutanal (*44*) [20] and the optically active steroidal C^{22}-aldehyde *47* [27] are two examples. Steroidal side chain extensions are often quite selective using RMgX [28], but titanation generally leads to improvements.

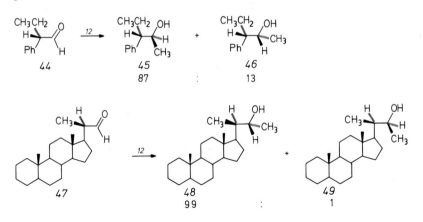

   Not much is known concerning the addition of titanium reagents to α-chiral ketones devoid of additional functionality in the vicinity of the carbonyl group. One example involves addition reactions of pregnenolone acetate (*50*) [8a, 20]. Previously, it had been reacted with CD_3MgI (*53*) to provide an 88:12 mixture of the (20S)- and (20R)-alcohols (*51a* and *52a*, respectively) [29]. 1,2-Asymmetric induction increases considerably by the use of the deuterated titanium reagent *54*, since only a trace of the (20R)-alcohol *52a* is formed. Titanation also helps in case of allyl-addition; the ^{13}C-NMR spectrum of the crude product shows essentially only a single diastereomer *51b* [20]. A clear limitation of organotitanium chemistry has to do with the fact that the less reactive *n*-alkyl analogs do not add to such sterically hindered ketones as *50* [20].

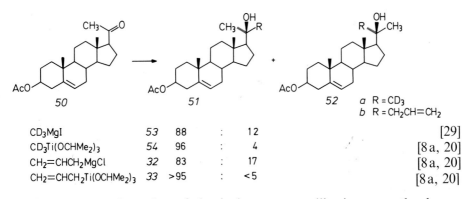

CD₃MgI		*53*	88	:	12	[29]
CD₃Ti(OCHMe₂)₃		*54*	96	:	4	[8a, 20]
CH₂=CHCH₂MgCl		*32*	83	:	17	[8a, 20]
CH₂=CHCH₂Ti(OCHMe₂)₃		*33*	>95	:	<5	[8a, 20]

In summary, titanation of classical organometallics increases the degree of Cram-preference in reactions with α-chiral aldehydes and ketones. Nevertheless, more work is necessary. A different and more difficult problem is the selective formation of anti-Cram products. The principle of double stereodifferentiation may turn out to be a viable solution [14, 30]. Reagent specific processes need to be developed [30]. So far, chirally modified titanium reagents (Section 5.5) have not been reacted with chiral aldehydes.

Chiral α-chloro aldehydes are known to react with Grignard reagents stereoselectively to form the "Cornforth products" preferentially, diastereomer ratios of ~6:1 being common [12]. Racemic *55* was chosen as a model system to study the effect of titanation. No significant differences were observed, although systematic variation of ligands or carbon nucleophiles remains to be carried out [27].

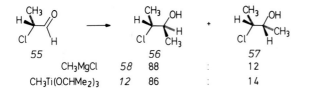

CH₃MgCl	*58*	88	:	12	
CH₃Ti(OCHMe₂)₃	*12*	86	:	14	

All of the above examples involve carbonyl compounds having a center of chirality. Diastereoselectivity is also possible in case of axially chiral aldehydes. The first such example pertains to *59*, which was reacted in racemic form with CH₃Ti(OCHMe₂)₃ (*12*) [31]. A mixture of diastereomers *60* was formed in a ratio of 68:32. The relative configuration remains to be established. However, the results show that the two diastereotopic π-faces of the aldehyde function are sterically similar and/or different conformations are actually reacting.

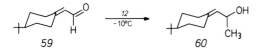

5.2.2 Chelation-Control in Addition Reactions of Chiral Alkoxy Carbonyl Compounds

Since Cram's original papers on chelation-controlled additions to chiral alkoxy and hydroxy ketones [4, 13], a great deal of progress has been made, titanium often being involved. In scrutinizing this area, it is useful to first consider various types of chelates systematically [32]. Organometallic reagents are potentially capable of forming chelates 62, 64, 66 and 68/69 from α-chiral α-alkoxy, α-chiral β-alkoxy, β-chiral β-alkoxy and α-chiral α,β-dialkoxy carbonyl compounds 61, 63, 65 and 67, respectively. Nucleophilic attack should then occur from the less hindered side as indicated by the arrows.

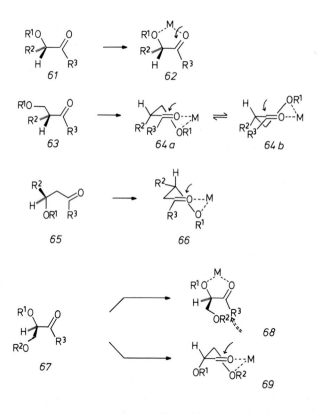

The problem is to find the right type of reagent for each situation. Work published up to 1980/81 can be summarized as follows [32]:

1) Grignard reagents react with ketones 61 (R^3 = alkyl, aryl) in THF with efficient chelation-control [13, 33]; alkyllithium reagents react less selectively.
2) The analogous aldehydes 61 (R^3 = H) generally fail to show efficient chelation-control in reactions with $RMgX$, RLi, R_2CuLi or other organometallics [13, 33].

3) Cuprates R_2CuLi, but not RLi or RMgX add stereoselectively to aldehydes *63* (R^3 = H) [34] to form chelation-controlled products. Allylzinc and tin reagents add to aldehydes related to *63* having additional chiral centers to form chelation-controlled products [35].

4) No systematic studies of addition reactions involving ketones *63* (R^3 = alkyl, aryl) are known.

5) In case of aldehydes *65* (R^2 = H) or ketones *65* (R^3 = alkyl), the use of RMgX, RLi, R_2CuLi or boron compounds fails to lead to acceptable levels of 1,3 asymmetric induction [13, 34].

6) Methods for chelation-controlled aldol additions to aldehydes *61* (R^3 = H) are unknown; for example, lithium enolates deliver mixtures in which the non-chelation-controlled products sometimes dominate slightly [14].

7) Certain lithium enolates add to aldehydes *63* (R^3 = H) with slight chelation-control (3:1 product ratios; improvements are possible only if a second chiral center is located at the β-position [14c].

8) Methods for chelation-controlled aldol additions to aldehydes *65* (R^3 = H) are unknown.

9) General methodologies for reversing diastereoselectivity, i.e., for non-chelation-control in Grignard or aldol additions to the above aldehydes are also not available.

10) Carbohydrates (e.g., *67*) are not included in the above points and often involve additional electronic and steric factors which must be considered in each particular case. A number of chelation-controlled reactions are known [36], but generalities as to which reagents are optimal cannot be made.

In recent times several major problems have been solved, particularly those mentioned in points 2), 5), 6), 7), 8) and 9). A review covering these developments up to early 1984 as well as other aspects of chelation and non-chelation has appeared [32].

5.2.2.1 1,2-Asymmetric Induction

With few exceptions [33, 37], α-alkoxy aldehydes of the type *61* (R^3 = H) do not react diastereoselectively with RMgX, RLi or other reagents such as allyl-boron compounds [33, 38]. Efficient chelation-control is possible in a general way by using Lewis acidic titanium reagents (Chapter 2). Thus, CH_3TiCl_3 (*9*) reacts with *70* to form a 92:8 product ratio of *72* and *73*, respectively [39]. Presumably, the octahedral chelate *71* is a short-lived intermediate which results in intra- or intermolecular transfer of the methyl group onto the sterically less hindered face of the aldehyde function. For comparison, CH_3MgI (THF/—30 °C), CH_3Li (THF/—78 °C) deliver *72:73* ratios of 60:40 and 40:60, respectively [39]. CH_3MgCl (THF/—110 °C) is more selective if a different protective group is used, but this method is not general [34].

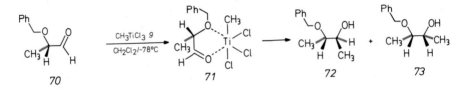

Since analogs of the type RTiCl$_3$ (R = n-alkyl, allyl) are usually unstable and/or diffucult to handle (Chapter 2), direct extension of the above method is not feasible. However, a variation provides a way out of the dilemma. Thus, 70 can be "tied up" by TiCl$_4$ to form 74, which then reacts stereoselectively with mild C-nucleophiles such as dialkylzinc, allylsilanes or allylstannanes [39] in methylene chloride. Ether or THF should be avoided because they destroy the chelate.

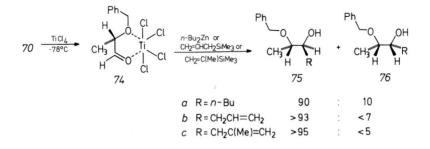

a	R = n-Bu	90 : 10
b	R = CH$_2$CH=CH$_2$	>93 : <7
c	R = CH$_2$C(Me)=CH$_2$	>95 : <5

The mechanism of C—C bond formation is not entirely clear. Either the Zn or Si reagents add directly to 74, or they first transfer the carbon nucleophile onto titanium, forming new intermediates similar to 71. It is interesting to note that non-complexed TiCl$_4$ rapidly reacts with R$_2$Zn in CH$_2$Cl$_2$ to form RTiCl$_3$, but CH$_2$=CHCH$_2$SiMe$_3$ does not lead to CH$_2$=CHCH$_2$TiCl$_3$ under similar conditions [31].

Although these mechanistic uncertainties have not yet been cleared up, it was possible to obtain direct NMR evidence for the intermediacy of such TiCl$_4$-complexes as 74. Since 74 begins to decompose at temperatures above −50 °C, the ^{1}H-NMR spectrum was recorded at −78 °C (Fig. 1). It shows a single species [27], in line with chelate 74. The position of the α-protons of the ether moiety is shifted downfield as expected, but the aldehyde proton signal hardly shifts relative to that of 70. The latter phenomenon is also observed for TiCl$_4$ complexes of normal aldehydes (e.g., of 23) [27]. The "non-equivalence" of the diastereotopic benzyl protons increases in going from the aldehyde 70 to chelate 74. The spectrum does not allow a decision as to the geometry around the ether function (which has "oxonium" character). In case of non-planarity, the oxygen is chiral and the benzyl group can be cis or trans to the neighboring H-atom.

Since the benzyl protective group works well in the above chelation-controlled reactions, not many other groups were tested [31]. It is interesting to note that the t-butyldimethylsilyl analog of 70 reacts with CH$_3$TiCl$_3$ to form a 17:83 mixture of chelation- and non-chelation-controlled adducts [31]. Similarly, treatment with $TiCl_4$ followed by the addition of $(CH_3)_2Zn$ results in a 26:74 product mixture, in which the non-chelation-controlled product again dominates. Apparently, the bulky t-butyldimethylsilyl group prevents chelation. General methods for non-chelation-control are discussed in Section 5.2.3.

In case of allylsilane (or the more expensive stannane) additions, $SnCl_4$ is equally well suited as a chelating and aldehyde-activating Lewis acid [39, 40]. $TiCl_4$ and $SnCl_4$ are similar in that both are capable of forming six-coordinate octahedral complexes with donor molecules. $MgBr_2$-etherate in CH_2Cl_2 (-30 °C/3 h) also promotes this reaction, but stereoselectivity (70:30) and conversion ($\sim 50\%$) are inferior [32]. One equivalent of Al_2Cl_6 ($AlCl_3$ is dimeric in solution) leads to an 85:15 ratio of $75b:76b$ (CH_2Cl_2/

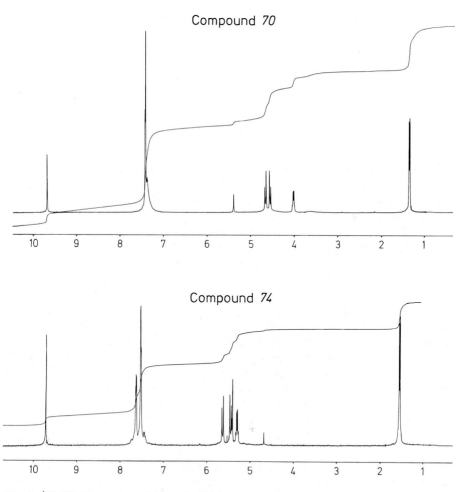

Fig. 1. ^{1}H-NMR spectra of *70* and of *74*.

−78 °C/3 h; ∼85 % conversion) [32]. Allylmagnesium chloride adds to *70* to form a 60:40 diastereomer ratio in favor of *75b* [39]. In contrast, BF_3 is incapable of chelation and actually induces reversal of diastereoselectivity [41] (Section 5.2.3). Crotylstannanes also add to chiral α-alkoxy aldehydes in the presence of $TiCl_4$ and other Lewis acids [42].

Since enol silanes were known to undergo Lewis acid promoted aldol additions to aldehydes [43], it seemed worthwhile to test them with titanium chelates such as *74*. Indeed, almost complete chelation-control and >90 % conversion was observed in all cases [39]. Sometimes the use of $SnCl_4$ results in slightly higher stereoselectivities [44].

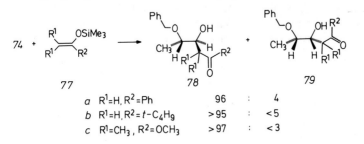

		78		79
a	R¹=H, R²=Ph	96	:	4
b	R¹=H, R²=t-C₄H₉	>95	:	<5
c	R¹=CH₃, R²=OCH₃	>97	:	<3

This is synthetically important, because it is the only currently known method for chelation-controlled aldol addition [44]. It was therefore of interest to determine how prochiral enol silanes behave. Since normal aldehydes are known to react fairly nonselectively [43] (e.g., *80* → *82* + *83*) [45], additions to *74* might be expected to deliver two of the four possible diastereomers.

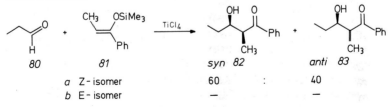

		syn 82		anti 83
a	Z-isomer	60	:	40
b	E-isomer	−		−

It was thus surprising that *74* reacts with *81a* to form essentially a single diastereomer *84* [39]. The additional stereochemistry (simple diastereoselectivity) is syn. The configuration of *84* was established by chemical correlation and an X-ray-structure determination [39, 44]. The use of $SnCl_4$ results in a similar stereoselection [39]. Interestingly, the E-isomer *81b* leads to similar results. Thus, to a first approximation, stereoselectivity is independent of the geometry of the enolate.

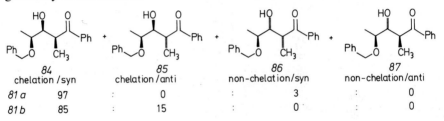

	84 chelation/syn		85 chelation/anti		86 non-chelation/syn		87 non-chelation/anti
81a	97	:	0	:	3	:	0
81b	85	:	15	:	0	:	0

The results do not prove any one mechanism [44], but are in line with an acyclic transition state *88* (for the Z-enolate) and *89* (for the E-enolate) in which the methyl group of the incoming nucleophile avoids steric interaction with the five-membered chelate [44].

In case of normal aldehydes, such an approach is no longer preferred due to the steric interaction between the methyl group and TiCl$_4$ (see *90* in case of the Z-enolate). Thus, the low degree of simple diastereoselectivity is due to the fact that TiCl$_4$ complexes in an anti-manner (*90*), while in chelates *74* syn-complexation pertains [44].

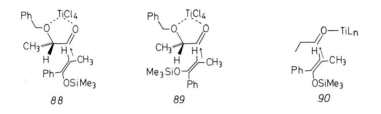

In view of the above, achiral, *91* should chelate via *92* and then lead to pronounced simple diastereoselectivity, which is indeed observed [44].

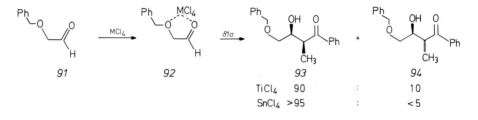

Further experiments show that the above reactions do not proceed via prior Si—Ti or Si—Sn exchange [44]. It is known that Z-configurated enol silanes react stereospecifically with TiCl$_4$ to form Z-configurated Cl$_3$Ti-enolates, which afford syn aldol adducts with aldehydes [46]. Upon treating *81a* with TiCl$_4$ and then adding *70*, a product ration (84:85:86:87 = 89:3:0:8) resulted which is different from the one previously observed for *74/81a* [44].

Various other prochiral enol silanes derived from ketones and esters were also tested. Whereas chelation-control is generally excellent, the degree of simple diastereoselectivity in favor of the syn adducts varies [32, 44]. This shows that not only the methyl group of the enolate, but also the bulk of the other enolate substituents exerts steric effects. In fact, the usual syn-selectivity is sometimes reversed [32]. The pure Z-enol silane from 3-penta-none adds to *74* with complete chelation-control, simple diastereoselectivity being anti (anti:syn = 82:18). A 17:83 mixture of Z/E enol silane results in complete chelation-control, but simple diastereoselectivity is different (anti: syn = 28:72) [31]. This underlines the mechanistic complexity in these reactions. Synclinical attack rather than *88* may be involved.

However, if the ketone is sterically hindered, intermediate chelates, e.g., *118*, analogous to *114* appear not to be involved, because the non-chelation-controlled product *120* is formed [21]. Instead, the Anh model *119* is in line with the results. The short Ti—O bond (1.75 Å; Chapter 2) would bring the substituents at the ring and the ligands at titanium in *118* in conflict. In line with this hypothesis is the observation that CH_3Li, CH_3MgCl and $CH_3Zr(OC_3H_7)_3$ react via chelation-control; $120:121 = 15:85$, $28:72$ and $19:81$, respectively [21]. The metal-oxygen bond is longer (1.9–2.1 Å) in all of these cases, making non-bonded interactions in the respective chelates less severe [8a, 21].

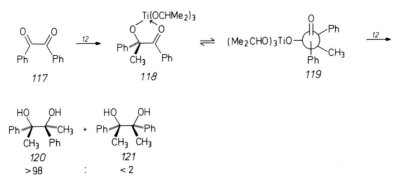

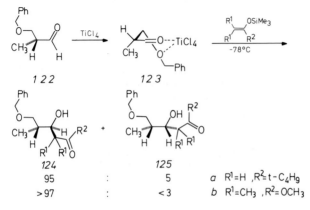

If the alkoxy group in α-chiral aldehydes is located at the β-position as in *63*, 1,2-asymmetric induction via chelation-control rests upon the intermediacy of six-membered chelates *64a/64b*. Since certain cuprates allow for such diastereoselectivity [34], efforts concentrated on aldol reactions [41]. Using the same strategy as in case of α-alkoxy aldehydes, β-chelate *123* from *122* was treated with enol silanes. In all cases excellent chelation-control was observed [41]. Prochiral enol silanes such as *81a* add with complete chelation-control, simple diastereoselectivity being syn [32, 48]. Here again chelation not only determines the sense of diastereofacial selectivity, but also that of simple diastereoselectivity. Indeed, achiral *126* reacts via *127* with excellent simple diastereoselectivity. In all of these cases, the use of $SnCl_4$ leads to lower diastereoselectivities [32].

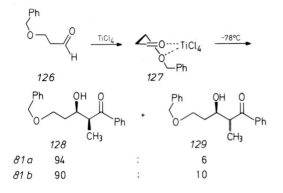

81 a   94   :   6
81 b   90   :   10

A number of organometallic reagents have been added to carbohydrate aldehydes and ketones with varying degrees of chelation-control [32, 36]. Unfortunately, general guidelines as to the optimum reagent are not available. Due to the presence of more than one protected hydroxy function, it is often difficult to predict whether α- or β-chelation dominates (67 → 68 versus 67 → 69). In case of β-chelation 69, the α-alkoxy group may exert a synergistic effect, i.e., it makes an "Anh-effect" possible. On the other hand, the latter effect (or the Cornforth model) may be the sole factor involved. It is therefore conceivable that reagents incapable of any chelation formally result in β-chelation-controlled diastereoselectivity. On the basis of a single reaction it is usually not possible to distinguish between these mechanistic alternatives [32]. A case in point is 2,3-O-isopropylidene-D-glyceraldehyde (130), prepared from mannitol. Although classical Grignard and enolate additions are not very selective, careful choice of reagents and conditions often allows for acceptable results [32]. Generally, diastereofacial selectivity is anti, which can be explained on the basis of the Anh- or Cornforth model (131 or 132, respectively), or by assuming β-complexation 133. The latter has been postulated for certain ZnX$_2$-mediated Grignard additions [52]. This is unlikely in case of reagents incapable of bisligation, non-chelation-control being a better explanation. These reactions are discussed in Section 5.2.3.

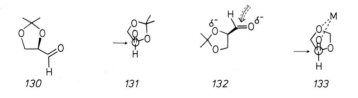

Since the acetonide 130 does not undergo α-chelation, it is not possible to obtain the syn-adducts. A recent solution to this problem involves the use of a differently protected form of D-glyceraldehyde, e.g., 134, which is also available from mannitol [48, 53]. With TiCl$_4$ it undergoes preferential α-chelation at the benzyloxy group according to 135. This allows for stereoselective addition of variety of carbon nucleophiles. *134* is an ideal building block because the products *136* have two different protective groups [48, 53].

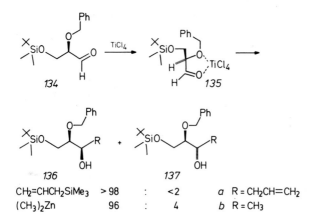

CH$_2$=CHCH$_2$SiMe$_3$	> 98	:	<2
(CH$_3$)$_2$Zn	96	:	4

a R = CH$_2$CH=CH$_2$
b R = CH$_3$

Turning from aldehydes and ketones to α-alkoxy carboxylic acid cyanides, the quantitative conversion *138* → *140* has been reported recently [49]. Stereoselective formation of *140* shows that the assumption of *139* as an intermediate is reasonable, i.e., alternative complexation at the cyano function does not occur.

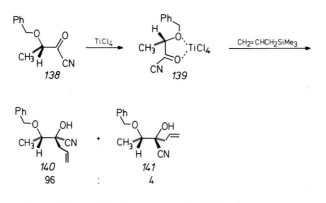

140		141
96	:	4

5.2.2.2 1,3- and 1,4-Asymmetric Induction

Control of 1,3-asymmetric induction in addition reactions of β-chiral β-alkoxy aldehydes *142* is not possible using such reagents as RMgX, RLi, R$_2$CuLi, lithium enolates or allylboron compounds [32]. Presently, the only way to solve this long-pending problem is to use Lewis acidic titanium reagents [45], e.g., CH$_3$TiCl$_3$ (*9*).

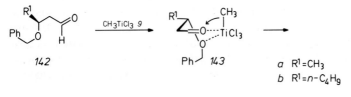

a R^1=CH$_3$
b R^1=n-C$_4$H$_9$

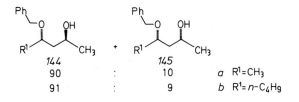

144	145	
90	10	a R¹=CH₃
91	9	b R¹=n-C₄H₉

Complexation of *142a–b* with $TiCl_4$ followed by the addition of dibutyl-zinc or allylsilanes results in 90–99 % chelation-controlled C—C bond formation [45]. Since the allyl group can be cleaved by ozonolysis, iterative additions are possible [32]:

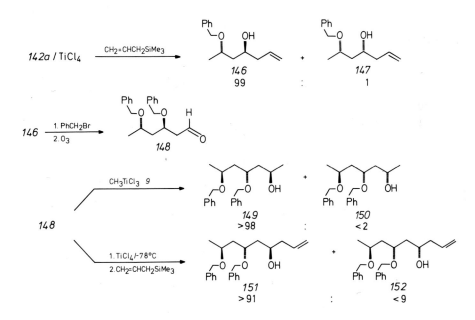

The reaction of *142a*/$TiCl_4$ with enol silanes also results in excellent diastereoselectivity [45]. In fact, it constitutes the currently only known method to perform aldol additions with chelation-controlled 1,3-asymmetric induction. Prochiral enol silanes such as *81a* add with almost complete chelation-control, simple diastereoselectivity being syn [45]. $SnCl_4$ results in mixtures.

Intramolecular transfer of C-nucleophiles [21, 45, 54], such as *158 → 160* [45] are beginning to be considered. Although it is not certain whether all of the reaction proceeds via initial deprotonation to form the Zr—O σ-bonded species *158*, intramolecular transfer of the allyl group should occur via the [1,3,3]bicyclic transition state *159* leading to preferential formation of *161*. This is indeed observed, which means that the sense of diastereofacial selectivity is the same as in the $TiCl_4$-mediated allylsilane addition to *142*. Attempts to observe related reactions using titanium reagents failed [55].

5. Stereoselectivity in the Addition of Organotitanium Reagents

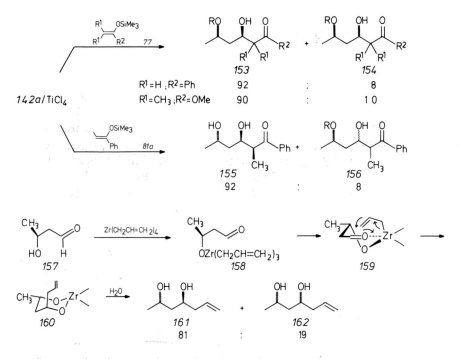

| | R¹=H ;R²=Ph | 92 | : | 8 |
| **142a/TiCl₄** | R¹=CH₃;R²=OMe | 90 | : | 10 |

Intramolecular reactions of such species as *158* are different from those of octahedral intermediates *143*, because the coordination number is different. An intermolecular process involving two species *143* would also explain the observed stereoselectivity. A completely different approach makes use of stable (allyl)siloxy aldehydes which react intramolecularly upon TiCl₄-activation [55, 56]. An example is *165* → *162* [56]:

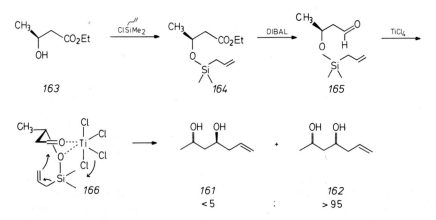

Intermediates of the kind *166* are related to the benzyl-protected analogs *142*/TiCl₄. In contrast to the latter, *166* cannot be isolated because complexation triggers immediate transfer of the allyl moiety from a direction as dictated by the spatial position of the silyl group. Synthetically, this concept

142

is valuable because it means inversion of diastereofacial selectivity relative to the previously attained 1,3-asymmetric induction using $142/TiCl_4/CH_2=$ $=CHCH_2SiMe_3$ or the intramolecular variation via the zirconium intermediate 158. In a formal sense, non-chelation-controlled products result. Previous attempts to achieve this end had failed; titanium reagents of low Lewis acidity such as $CH_2=CHCH_2TiX_3$ (X = $OCHMe_2$, NEt_2) react non-selectively with 142, as do RLi, RMgX and boron reagents [45].

Returning to intermolecular chelation-controlled C—C bond formation, acid nitriles such as 167 can be used in place of aldehydes 142 [49]. 1,3-Asymmetric induction is >99%. The tentative configurational assignments are based on the assumption of carbonyl chelation 168. Remarkably, the aldol addition using $81a$ afford only one of four possible diastereomers (171). This is another example of the influence of chelation on simple diastereoselectivity [49].

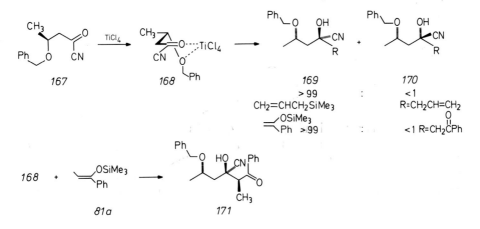

167   168   169   170

169 : 170
> 99 : < 1
$CH_2=CHCH_2SiMe_3$ R=$CH_2CH=CH_2$
OSiMe₃
Ph >99 : <1 R=$CH_2\overset{O}{\overset{\|}{C}}Ph$

168 + 81a → 171

Only a few cases of appreciable 1,4-asymmetric induction in carbonyl addition reactions are known [21, 39]. In case of 172 a more flexible seven-membered $TiCl_4$ chelate is likely to be involved [39], resulting in somewhat lower degrees of chelation-control relative to those in the 1,3-system. Nevertheless, Lewis acidic titanium reagents are currently the only compounds which make such 1,4-asymmetric induction possible. It is likely that chirally modified reagents will solve such problems in a more general way (reagent-control).

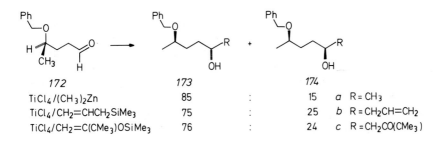

172   173   174

	173		174	
$TiCl_4/(CH_3)_2Zn$	85	:	15	a R= CH_3
$TiCl_4/CH_2=CHCH_2SiMe_3$	75	:	25	b R = $CH_2CH=CH_2$
$TiCl_4/CH_2=C(CMe_3)OSiMe_3$	76	:	24	c R = $CH_2CO(CMe_3)$

143

1,4-Asymmetric induction was also observed in the double addition to the dialdehyde *175* [21]. Whereas CH$_3$MgI yields a 50:50 mixture of *177* and *178*, two equivalents of CH$_3$Ti(OCHMe$_2$)$_3$ (*12*) result in appreciable 1,4-asymmetric induction, in line with the chelate *176* [21]. In contrast, Ti(CH$_4$)$_4$ reacts rather unselectively (*177:178* = 42:58) [21]; it is possible that the transfer of the second methyl group occurs to some extent intra-molecularly via a [1,1,4]bicyclic transition state. In line with this speculation is the observation that a six-fold dilution in case of the Ti(CH$_3$)$_4$ reaction reverses diastereoselectivity (*177:178* = 52:48) [21].

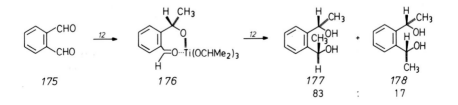

175   176   177   178
                                          83   :   17

5.2.3 Non-Chelation-Controlled Additions to α-Chiral Alkoxy Carbonyl Compounds

Non-chelation-control is a difficult task because there is no general way to reduce the degrees of freedom of non-complexed molecules. Reagents in-capable of chelation must be used and electronic and/or steric factors relied upon, notably those defined by the Felkin-Anh or Cornforth (dipolar) models. Although the method of TiCl$_4$-induced intramolecular allyl transfer in β-chiral β-(allyl)siloxy aldehydes results in complete reversal of 1,3-asymmetric induction and thus simulates "non-chelation-control" (Section 5.2.2.2), this strategy is not expected to be successful in case of α-(allyl)siloxy analogs [56]. Therefore, new strategies had to be developed.

Since Lewis acidity of alkyltitanium reagents decreases drastically in going from RTiCl$_3$ to RTi(OR')$_3$ (Chapter 2), the latter could be expected to be incapable of chelation. α-Alkoxy aldehydes might then react via non-chelation due to the electronic effects inherent in the Felkin-Anh or Cornforth models. Indeed, upon adding CH$_3$Ti(OCHMe$_2$)$_3$ (*12*) to *70*, the Felkin-Anh product *73* was formed preferentially [8a, 39]. Thus, switching from CH$_3$TiCl$_3$ to CH$_3$Ti(OCHMe$_2$)$_3$ reverses diastereofacial selectivity. The acetate *179* reacts chemo- and stereoselectively [39].

This methodology can be applied to carbohydrate chemistry. Whereas the furanose *182* reacts with CH$_3$MgBr under chelation-control [57], CH$_3$Ti(OCHMe$_2$)$_3$ results in the opposite diastereoselectivity [8a, 39]. Another example is the protected triose *134* which reacts with CH$_3$Ti-(OCHMe$_2$)$_3$ under non-chelation-control [48, 53], in complete contrast to TiCl$_4$/(CH$_3$)$_2$Zn (Section 5.2.2.1).

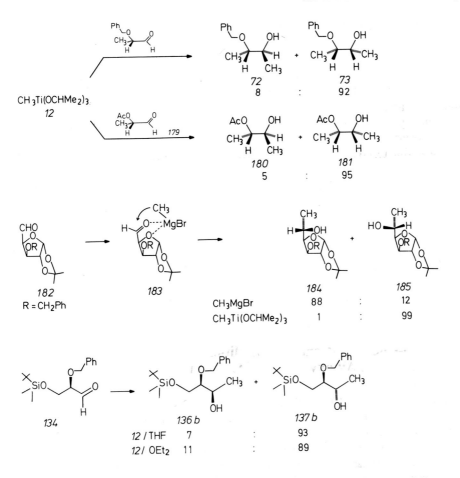

In order to achieve non-chelation-controlled aldol additions, triisopropoxytitanium enolates can be used [39, 41, 44]. Just like $CH_3Ti(OCHMe_2)_3$, they are reagents of low Lewis acidity, incapable of efficiently "tying up" aldehydes such as 70. The prochiral enolate 186 leads to two of four diastereomers [39, 41].

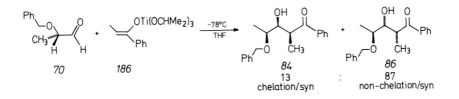

Sometimes tris(diethylamino)titanium enolates (made by reacting Li-enolates with $ClTi(NEt_2)_3$ (188) [26, 48]) are better suited [32]. For example, 189 adds to 70 with 93 % non-chelation-control and > 90 % simple diastereoselectivity (syn) [32, 48]:

5. Stereoselectivity in the Addition of Organotitanium Reagents

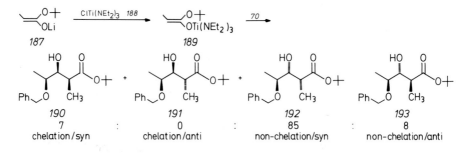

190 : 191 : 192 : 193

7 : 0 : 85 : 8

chelation/syn : chelation/anti : non-chelation/syn : non-chelation/anti

Alternative strategies designed to achieve non-chelation-controlled aldol additions to α-alkoxy aldehydes include [32]:

1) $FN(C_4H_9)_4$ induced additions of enol silanes [41];
2) Addition of enol silanes to doubly BF_3-complexed aldehydes 70 [41];
3) Application of the principle of chirally modified reagent-specific compounds.

So far, none of these methods consistently lead to high levels of non-chelation-control. This also applies to the above titanium enolates although reagent control using optically active alkoxy or amine ligands has not been explored to date.

Simple diastereoselectivity in the reaction of titanated heterocycles with aldehydes was first reported in 1982 (Section 5.3.2) [8a]. An example in which simple diastereoselectivity and diastereofacial selectivity is relevant involves the titanated bis-lactim ether 195 [58]. Although it has been formulated as a pentahapto species 195A [58], the σ-bonded form 195B appears to be more likely. In any case, addition to the mem-protected form of S-lactaldehyde 196 affords essentially only one of four diastereomers (197). This means non-chelation-control and excellent simple diastereoselectivity (Section 5.3.2) Acylation of the hydroxy group in 197 followed by hydrolysis of the heterocycle and esterification provides pure 198 [58].

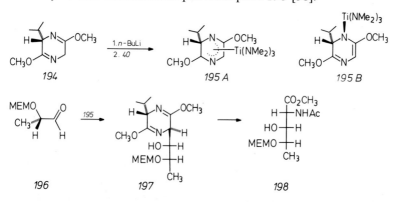

Another application of non-chelating titanium reagents [8a, 32, 39, 44, 53] concerns the first example of a diastereoselective homo-aldol addition; 39 reacts with 70 to produce two of eight possible diastereomers [25].

146

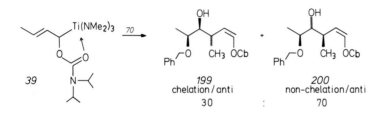

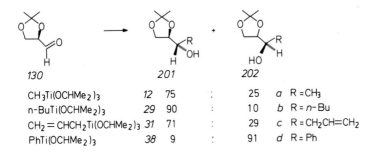

As mentioned in Section 5.2.2.1, glyceraldehyde-acetonide *130* and similar compounds react with certain organometallic reagents to form anti-adducts preferentially. This trend can be explained by assuming either β-chelation or no chelation at all, in which case the Felkin-Anh or Cornforth models operate. Since titanium reagents to low Lewis acidity do not show chelation effects in reactions with α- or β-alkoxy aldehydes, they are not expected to chelate in systems of the type *130*. Indeed, $CH_3Ti(OCHMe_2)_3$ (*12*) adds to *130* to produce a 75:25 mixture of anti/syn adducts *201a* and *202a*, respectively [27, 32]. Although there is room for improvement, this is currently the best way to introduce methyl groups stereoselectively (the Grignard reagent delivers poor results) [32]. The same sense of diastereofacial selectivity is observed in the analogous addition of *n*-butyl and allyl groups, but $PhTi(OCHMe_2)_3$ (*38*) results in preferential formation of the syn-adduct *202d* [59]. This is a rare exception in stereoselective additions to *130* and has not been explained to date. In case of allyl and crotyl additions to *130* and related sugars, zinc [59, 60] and boron reagents [61] have been used with great success. Crotylitanium reagents have not been reacted with *130*.

	130	*201*		*202*	
$CH_3Ti(OCHMe_2)_3$	12	75	:	25	*a* R = CH_3
n-BuTi$(OCHMe_2)_3$	29	90	:	10	*b* R = *n*-Bu
$CH_2 = CHCH_2Ti(OCHMe_2)_3$	31	71	:	29	*c* R = $CH_2CH=CH_2$
PhTi$(OCHMe_2)_3$	38	9	:	91	*d* R = Ph

Double diastereodifferentiation has been tested in the non-chelation-controlled addition of the optically active titanated heterocycle *195A* to R- and S-2,3-0-cyclohexylidene glyceraldehyde *203* and *206*, respectively [58]. The results show that double diastereodifferentiation does not play a major role [62]. Following the usual transformation of the initial adducts, stereochemical pure products *205* and *208* are accessible [58]. The Felkin-Anh model is in line with these results. The tris(diethylamino) analog of *195A* reacts similarly, but the lithium precursor adds less selectively [58].

147

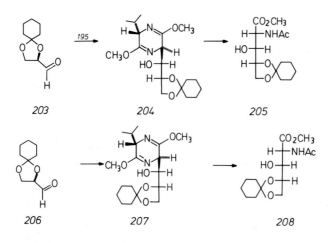

203   204   205

206   207   208

Although titanation of carbanions with $ClTi(OCHMe_2)_3$ or $ClTi(NR_2)_3$ generates reagents which are well suited for non-chelation-controlled additions, this methodology does not extend to α-alkoxy ketones. For example, $CH_3Ti(OCHMe_2)_3$ (*12*) adds to 2-benzyloxy-3-pentanone with complete chelation-control [31] (Section 5.2.2.1). The more Lewis basic ketones undergo chelation even with reagent *12*. Thus, the problem of non-chelation-controlled additions to such chiral ketones is particularly difficult. It was finally solved by replacing the benzyl by silyl protective groups. Thus, *209* reacts with $CH_3Ti(OCHMe_2)_3$ (*12*) to form virtually a single (non-chelation-controlled) product *211* [63]!

Apparently, the bulky group prevents coordination with CH_3Ti-$(OCHMe_2)_3$, allowing the factors defined by the Felkin-Anh or Cornforth model to operate. Classical reagents such as CH_3Li or CH_3MgX afford mixtures. Titanium ester enolates also react with 100% non-chelation-control [63].

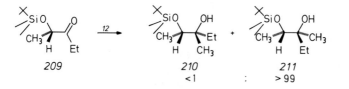

209   210   211
                 <1   :   >99

5.3 Simple Diastereoselectivity

The union of two prochiral centers of achiral molecules creates two centers of chirality. Simple diastereoselectivity in such processes forms the basis of most of the recent work on aldol reactions [14] and related reactions of crotylmetal reagents [15]. Although a number of special cases are known, the most general rule has been formulated as follows [14, 15]:

Z-configurated enolates or crotyl metal compounds form syn-adducts preferentially, while E-configurated reagents favour the anti-diastereomers (Masamune nomenclature).

These trends have been explained by assuming a six-membered cyclic transition state having a chair geometry. The rules of conformational analysis can then be applied, particularly if the transition state is tight. This is evident in the addition of E-crotylmetal compound *213*. Of the two chair transition states *214* and *217*, the former is energetically more favorable [15]. Assuming chair transition states are of lower energy than boat analogs *216/219*, it becomes clear why product *215* dominates over *218*. Analogous pictures for E-configurated reagents show that *218* should be the preferred diastereomer. The analysis of kinetically controlled ketone-enolate additions to aldehydes follows the same lines. However, the situation is more complicated because the sp^2-hybridized C-atom of the enolate closest to the metal bears an R-group instead of a small H-atom as in *213*. Because this R-group interacts sterically in the transition state, stereoselection in case of E-enolates is not as pronounced as in case of the Z-counterparts; a switch to boat transition states may occur [14].

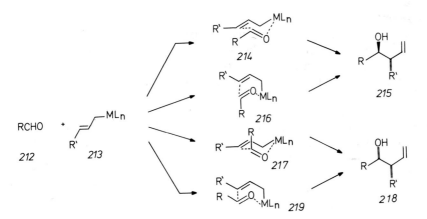

Boat transition states have been postulated even in case of certain substituted allylmetal reagents [15]. The similarity between the addition reactions discussed here and the Claisen rearrangement (in which boat-like transitions are only 2–3 kcal/mol (8–12 kJ/mol) less stable than the chair forms) suggests that such hypotheses may be reasonable [15]. Furthermore, stepwise mechanisms are possible, e.g., coordination of the metal to the aldehyde oxygen followed by rearrangement of the intermediates [64]. Acyclic transition states also need to be considered in some cases [65].

5.3.1 Titanium-Mediated Aldol Additions

The problem of simple diastereoselectivity in aldol additions has been largely solved by using prochirally pure or enriched enolates, the metal often being lithium or boron [14]. As noted above, there is room for improvement in case of E-enolates, because anti-selectivity is not uniformly acceptable. However, a number of special syn- and anti-selective enolates derived from certain ketones and esters are now available [14]. In such cases the initial adduct must be chemically modified in order to obtain compounds which are useful in further C—C bond formation. For example, the aldol adducts from certain ester enolates can be reacted with DIBAL, which reduces the ester function to the desired aldehyde. Formally, an aldehyde enolate has thus been added stereoselectively, which cannot be achieved directly.

Apart from mechanistic interest associated with stereoselective aldol reactions, titanium chemistry has turned out to be meaningful in the following three areas:

1) Syn-selective reactions of enolates derived from cyclic ketones.
2) Diastereoselectivity irrespective of the geometry of the enolate.
3) Diastereoselective additions of aldehyde enolates (or equivalents).

In order to perform stereoselective aldol additions, methodologies for the synthesis of prochirally pure metal enolates had to be developed [14]. Stereoconvergent aldol additions in which Z- and E-enolates show the same sense of simple diastereoselectivity would constitute an attractive alternative. This strategy was first realized in the reactions of bis(cyclopentadienyl)chloro-zirconium- [66a, b], triphenyltin [66c] and tris(dialkylaminosulfonium- [65a] (TAS) enolates. These systems yield synadducts preferentially, irrespective of the geometry of the enolate. In situ preparation of certain tin-enolates using Sn(II) reagents followed by aldol addition also results in syn-adducts, although the geometry of the enolates is unknown [66d].

Titanium enolates bearing alkoxy or amino ligands are best prepared by reacting the lithium analogs with the proper titanating agent [26]. Although the structure of these reagents has not been elucidated in the vast majority of cases, the enolate 222 was flash distilled (100–170 °C bath temperature/0.05 torr, 78% yield) and shown by NMR (Fig. 2) to have the O-titanated form; no sign of a C-titanated species was detected [8a, 67]. The analogous triisopropoxytitanium enolate 186 in thermally less stable and largely decomposes upon attempted distillation. This is reminiscent of the increased thermal stability of $RTi(NEt_2)_3$ relative to $RTi(OCHMe_2)_3$ (Chapter 2).

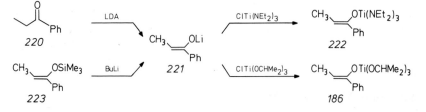

The vinyl proton of 222 absorbs at $\delta = 4.83$, compared to $\delta = 5.30$ of the corresponding signal of the enol silane 223 [67]. The upfield shift in going from Si to Ti is qualitatively similar to that observed previously for enol

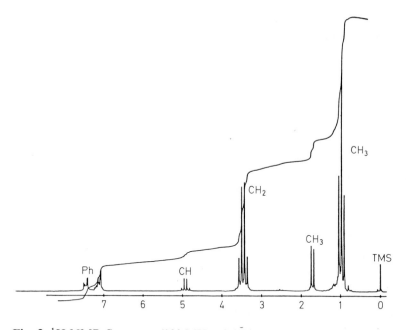

Fig. 2. ^{1}H-NMR Spectrum (100 MHz, CDCl$_3$) of the Enolate *222* [67]

silanes versus Li-, Na- and K-enolates [68]. The ^{13}C-NMR spectrum (CDCl$_3$) of *222* shows vinyl carbon absorptions at $\delta = 98.1$ and 159.1 (the latter being assigned to the carbon atom bearing the oxygen). These are different from the corresponding values observed for the enol silane *223*: $\delta = 105.2$ and 149.9, respectively [67]. In particular, upon going from *223* to *222*, the signal of the α-carbon (bearing the methyl group) shifts upfield by 7.1 ppm. In case of Li-, Na- and K-enolates, this shift amounts to 13–16 ppm, and has been interpreted on the basis of increased charge at α-C-atom [68]. Using the corresponding enol acetate as a standard, the following upfield shifts evolve:

Enol acetate	Enol silane	Ti-Enolate	Li-, Na-, K-Enolates
$\Delta\delta$(ppm) 0	8	15	20–24

Within this order, (Et$_2$N)$_3$Ti-enolates take on a position between the Si- and Li-analogs. Finally, the CH coupling constant at the α-position amounts to J$_{CH}$ = 155 Hz in case of *222*, which is similar to that observed for enol silanes and acetates, in line with sp^2-hybridization [67].

(Et$_2$N)$_3$Ti- and (Me$_2$CHO)$_3$Ti-enolates are red-brown (sometimes yellow-orange), air sensitive species which are soluble in all common non-protic solvents (including pentane). Usually, even the triisopropoxy derivatives are so stable that the solvent can be stripped off at room temperature and a different solvent added if desired.

Very little is known concerning aldol additions of aldehyde-enolates with aldehydes [14]. In fact, several lithium enolates derived from aldehydes

5. Stereoselectivity in the Addition of Organotitanium Reagents

were shown to react essentially stereorandomly, e.g., *224* → *225/226* [14]. Another problem concerns the low stability of the aldols; thus, protection of the hydroxy function is often necessary, e.g., in the form of the acetate.

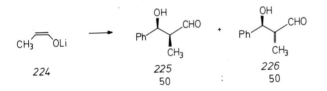

| | | 224 | | 225 | : | 226 |
| | | | | 50 | | 50 |

Titanium aldehyde-enolates were first prepared using *227*, which leads to a reagent *228* lacking prochirality [67]. Addition to benzaldehyde is smooth (>80% conversion), but the adduct *229* starts to decompose after standing at room temperature for several hours. Steglich-acylation provides stable *230* (61% isolated).

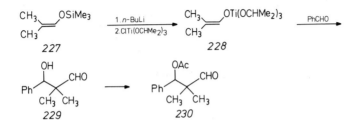

Unfortunately, the reactions of mono-substituted titanium aldehyde-enolates (e.g., *231*) with benzaldehyde are of limited value, because acylation leads to elimination products *233* [67].

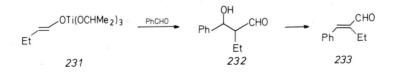

The situation is completely different in case of titanium ketone-enolates. The lithium percursors can be prepared either by LDA induced deprotonation of ketones (the amine present in the mixture has no effect on the aldol addition), or by desilylation of enol silanes using methyl- or *n*-butyllithium [67]. The following data (Table 1) concern reactions of titanium enolates derived from cyclic ketones with aldehydes [26, 67].

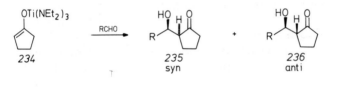

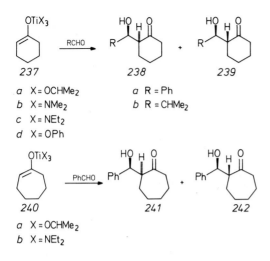

237

a X = OCHMe₂
b X = NMe₂
c X = NEt₂
d X = OPh

a R = Ph
b R = CHMe₂

238

239

240

a X = OCHMe₂
b X = NEt₂

241

242

Table 1. Aldol Additions of Titanium-Enolates Derived from Cyclic Ketones [26, 67]

Enolate[a]	R in RCHO	syn:anti	syn:anti in case of Li-Enolate [14]
234	Ph	85:15	—
234	CHMe₂	—	5:95
237a	Ph	86:14	52:48
237b	Ph	82:8	52:48
237c	Ph	93:7	52:48
237d	Ph	90:10	52:48
237a	CHMe₂	95:5	—
240a	Ph	91:9	—
240b	Ph	90:10	—

[a] Reactions were generally performed in THF at −78 °C, conversion being >90%.

Although systematic optimization with regard to ligands remains to be carried out, the data allow for some generalizations. In all cases the syn-adduct is formed preferentially. This is synthetically important because the Li- enolates are either anti-selective (as are boron enolates) [14] or deliver mixtures. Bis(cyclopentadienyl)chloro-zirconium enolates from cyclic ketones react with aldehydes to form syn-adducts with moderate syn-selectivity (72:28 syn/anti ratios) [66b], as do triphenyltin enolates [66c]. Other tin enolates (prepared in situ from cyclic ketones or their α-bromo-analogs) show excellent syn-selectivity (95:5 diastereomer ratios), but the yields are unacceptable (∼30%) [66d]. One case of a TAS-enolate from a cyclic ketone involves excellent syn-selectivity at short reaction times, but conversion is not optimal under such conditions; longer reaction times increases the yield at the expense of diastereoselectivity [65a]. Thus, reactions of titanium enolates derived from cyclic ketones constitute the currently best method for obtaining syn aldol adducts.

The data in Table 1 also show that switching from isopropoxy to amino groups can have a beneficial effect. Within the latter series, NEt$_2$ ligands are better suited than the NMe$_2$ analogs, probably due to steric factors. Apart from the synthetic side, the effect of ligand on diastereoselectivity is not readily compatible with an acyclic transition state 243 in which titanium is "far removed" from the reaction centers as shown in 243. On the other hand, a cyclic mechanism involving the traditional chair geometry leads to the wrong stereochemistry (anti-adducts). Thus, a boat transition state 245 may be viewed as an acceptable explanation, although the reason for this preference is unclear.

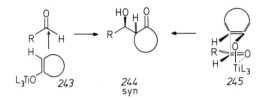

The above method was applied to the stereoselective side chain extension of steroids [21]. Compound 248 reacts with benzaldehyde to form essentially one of four possible diastereomers; the syn adduct 249 can be obtained in pure form after one recrystallization (66% isolated yield). In contrast, the lithium enolate 247 affords a 25:75 mixture of 249/250 in addition to other products [21].

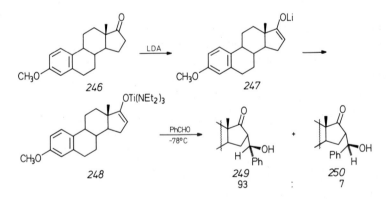

The addition of Li-enolates to Ti(OCHMe$_2$)$_4$ at low temperatures results in species which have been formulated as titanium ate complexes, although the structural data are not yet available [26]. This was the first case of titanium ate complexes in organic synthesis. Their chemical behavior is different from that of the lithium precursor or of the corresponding triisopropoxytitanium enolate. For example, 251 reacts with Ti(OCHMe$_2$)$_4$ to afford a yellow solution, whereas *237a* forms orange-red solutions. Also, the assumed ate complex *252* adds to benzaldehyde to form a 70:30 diastereomer ratio of *238 a/239 a*, which is different from the aldol addition of *251* (*238 a*: *239 a* = 48:52) and *237a* (*238a*:*239a* = 84:16). Although this could result from an equilibrium mixture *251* + Ti(OCHMe₂)₄ $\rightleftarrows$ *252* + *237a*, an ate complex *252* as the reacting species appears likely. The results of chemo- and regio-selective addition of similar species (ester enolates + Ti(OCHMe₂)₄) also point to the existence of ate complexes [67, 69, 70]. Since initial experiments with *252* led to lower diastereoselectivities than *237*, further studies were not carried out [26]. However, it is worth pointing out that allyltitanium ate complexes are very useful reagents [8 a].

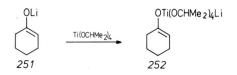

Turning to trialkoxy- or trisdialkylaminotitanium enolates derived from acyclic ketones, it was found that they form syn aldol adducts preferentially, irrespective of the geometry of the enolate, although exceptions occur [26] (Table 2).

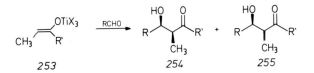

The results are not easily interpreted. A switch from boat to chair cyclic transition states appears to be operating, depending upon the enolate structure and the ligands at titanium. The triphenoxy titanium enolates are often most selective. However, they are not as readily available (ClTi(OPh)₃ has to be used) [67]. In summary, titanium enolates from acyclic ketones are of synthetic interest only in case of stereoconvergence, i.e., when mixtures of Z/E-enolates afford acceptable levels of diastereoselectivity. It should be mentioned that the reactions of the corresponding lithium enolates generally lead to poor results [14, 67].

A related approach makes use of trichlorotitanium enolates, prepared by the interaction of TiCl₄ with enol silanes [46, 71, 72]. A limitation of this method has to do with the fact that only Z-configurated enol silanes react with good yields. For example, the cyclopentanone derivative decomposes as it forms. An exception are trichlorotitanium enolates derived from cyclo-hexanones; they have been characterized by NMR spectroscopy. For example, the ¹³C-NMR signals of the olefinic carbon atoms of *257* appear at 181.1 and 114.8 [46], which is exceptional for an enol derivative [68]; in *256* they appear at δ = 157.0 and 101.4. It has thus been said that the trichlorotitanium

Table 2. Aldol Additions of Titanium Enolates *248* Derived from Acyclic Ketones [26, 27]

R^1	X	Z:E	R	syn:anti[a] (254:255)
Et	OCHMe$_2$	30:70	Ph	89:11
Et	NEt$_2$	30:70	Ph	76:24
Et	NMe$_2$	30:70	Ph	68:32
Et	OCHMe$_2$	30:70	c-C$_6$H$_{11}$	77:23
Et	OCHMe$_2$	30:70	t-C$_4$H$_9$	81:19
Et	OCHMe$_2$	66:34	Ph	85:15
Et	NEt$_2$	66:34	Ph	43:57
Et	OCHMe$_2$	92:8	Ph	88:12
Et	NEt$_2$	92:8	Ph	41:59
Ph	OCHMe$_2$	>98:<2	Ph	87:13
Ph	NEt$_2$	>98:<2	Ph	23:77
Ph	NMe$_2$	>98:<2	Ph	31:69
Ph	OPh	>98:<2	Ph	95:5
Ph	OCHMe$_2$	>98:<2	Et	89:11
Ph	OPh	>98:<2	Et	90:10

[a] All reactions at $-78\,°C$ (conversion $>85\%$); ratio determined by capillary gaschromatography or ^{13}C-NMR spectroscopy.

group exerts a significant electron withdrawing effect instead of an electron release usually expected for a metal atom. This is reminiscent of the low field ^{13}C-absorption of CH$_3$TiCl$_3$ (Chapter 2). The aggregation state (Chapter 2) of the enolates has not been ascertained. Generally, the trichlorotitanium enolates favor formation of syn aldol adducts (up to 89:11 ratios), although exceptions are known [46, 71]. Evidence for a boat transition state has been presented, but the origin of this preference again remains unclear. The ultimate cause may be secondary orbital interactions favoring the "endo" arrangement of reactants (cf. *245*) [46, 71]. Enol silanes react with SnCl$_4$ to form a α-SnCl$_3$ ketones, which react syn-selectively with aldehydes [71]. However, the yields are often poor.

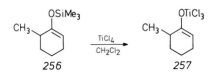

256 → 257

The aldol additions of simple lithium ester enolates usually fail to show pronounced diastereoselectivity [14]. Not much is known concerning the effect of titanation [67, 73b]. The aldol addition of the lithium enolate *258* (which contains 5% of the Z-isomer) to benzaldehyde at $-78\,°C$ results in a 62:38 mixture of *260* and *261*, respectively [14]. Adding a cooled solution of benzaldehyde to the titanium enolate *259* in THF at −78 °C results in an 87:13 ratio of *260*:*261* [67]. More work is needed in this area. *Bis*-silylketene ketals undergo TiCl₄-mediated syn-selective additions, although reversal of diastereoselectivity occurs sometimes [73a]. In case of the TiCl₄ and Cl₂Ti(OCHMe₂)₂ mediated aldol additions of S-silyl ketene S,N-ketals, a boatlike transition state has been postulated [73c].

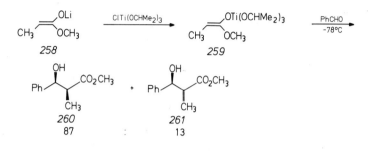

Titanation of Horner-Emmons reagents influences the reaction mode with aldehydes, depending upon the titanating agent [67]. The addition of Ti(OCHMe₂)₄ to the sodium salt *263* followed by condensation with benzaldehyde affords E-cinnamic acid ester *264*, just like *263* itself. However, titanation with ClTi(OCHMe₂)₃ results in a species which undergoes a completely stereoselective Knoevenagel condensation (22 °C/3 h) with the formation of the Z-isomer *265a* (57% yield). Under such conditions chemo-selective transesterification at the carboxylic acid ester function occurs. The same transformation is possible by stirring *262* with ClTi(OCHMe₂)₃ in the presence of triethylamine (78% yield of *265a*), but in this case stereoselectivity is not complete (~6% of *265b*). The results may be explained by stereoselective addition of a titanium enolate followed by stereospecific fragmentation

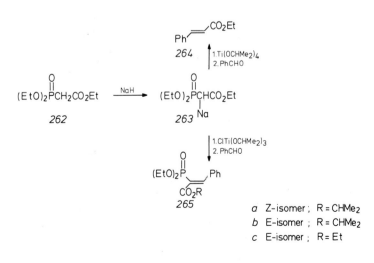

involving loss of $HOTi(OCHMe_2)_3$ instead of Horner-Emmons elimination. Other aldehydes react similarly [67]. In contrast, Knoevenagel condensation of *262* using $TiCl_4$/triethylamine results in the formation of the thermo-dynamically more stable E-isomer *265c* [74]. Knoevenagel condensations are also smooth in case of other active methylene compounds CH_2Y_2 ($Y = CO_2R$, CN, etc.) using the $ClTi(OCHMe_2)_3$/triethylamine [67] or $TiCl_4$/triethylamine system [74]; such conditions often give better yields than those of the classical Knoevenagel condensation.

Lithium and triisopropoxytitanium enolates derived from aldehydes do not react stereoselectively with aldehydes [14, 67]. Thus, a different strategy based on aldol-like addition of α-deprotonated aldehyde hydrazones was developed [75]. Whereas the lithium precursor *267* [76] do not react diastereoselectively with aldehydes [75, 76a] (with a few exceptions), tita-nation solves the problem. In particular, the triisopropoxy ligand system leads to excellent syn-selectivity (Table 3). The question of C- of N-titanation was investigated in one case using NMR spectroscopy. The ^1H-NMR spec-trum of *268d* (Fig. 3) is in line with the N-titanated form as shown, in which the configuration of the olefinic system is E (doublets at δ 5.8 and 7.5; $J = 14.0$ Hz) [77].

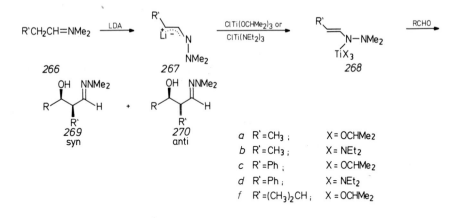

The results show that the triisopropoxy ligand system is better suited than the tris(diethylamino)analog. It should be noted that in the former case the reagents *268* (X = $OCHMe_2$) may undergo ligand exchange processes, as shown by NMR spectroscopy [75]. Thus, several reacting species are likely to be involved. The hydroxy group of the adducts can be protected using t-butyldimethylchlorosilane in the presence of four equivalents of imidazole

Table 3. Diastereoselective Addition of Titanated Aldehyde-Hydrazones to Aldehydes [75]

Ti-Species	R in RCHO	Conversion (%)	*269:270*
268a	Ph	80	91: 9
268b	Ph	61	85:15
268a	CH_3	61	95: 5
268a	CMe_3	70	93: 7
268c	Ph	95	98: 2
268c	p-NO_2—Ph	40	98: 2
268c	CH_3	95	96: 4
268d	CH_3	50	90:10
268e	Ph	78	94: 6

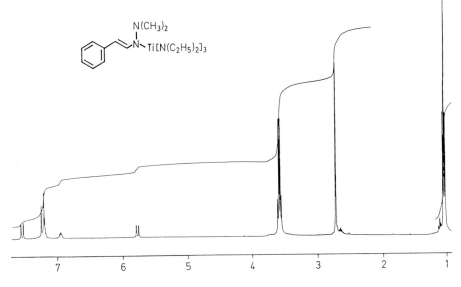

Fig. 3. ^{1}H-NMR Spectrum of *268d*.

(DMF/22 °C/60 h; 75% yield) [78]. Titanated ketone hydrazones also react syn-selectively [75]:

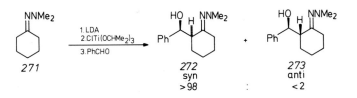

5. Stereoselectivity in the Addition of Organotitanium Reagents

In order to gain more information regarding the mechanism, the effect of varying the nature of the groups at the terminal nitrogen was studied [78]. Reagent *274* adds to benzaldehyde with essentially the same diastereoselectivity as the corresponding *N,N*-dimethyl analog *268a* (91:9 product ratio). Thus, the terminal nitrogen is not likely to be intimately involved in determing the sense and degree of stereoselection.

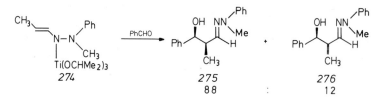

	274		275	276
			88	12

Since it was not possible to generate Z-configurated analogs of *268*, the mechanism of addition is a matter of speculation. Assuming E-configuration (as proven in case of *268d*), a boat transition state *277* formally accounts for the observed syn-selectivity. However, this does not readily explain the increase in diastereoselectivity as the size of R^1 in the titanium reagent increases (Table 3). An open transition state *278* (with or without assistance of LiCl) should also be considered.

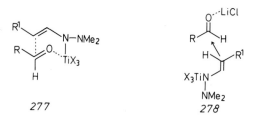

Finally, TiCl$_4$ mediated reactions of the *N*-silylated hydrazone *280* were tested. Preliminary results point to a switch in the sense of diastereoselectivity in going from benzaldehyde to pivalaldehyde [48]; more work is necessary in this area.

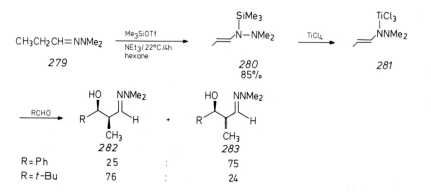

	282		283
R = Ph	25	:	75
R = t-Bu	76	:	24

5.3.2 Aldol-Type Additions of Titanated Heterocycles

Although titanated heterocycles have rarely been used for addition to aldehydes, this area seems promising. The first examples to be reported involve the titanated form of lactones and lactams, prepared from the lithium precursors [8a, 67]. Whereas the triisopropoxy ligand system results in syn/anti ratios of 40:60, the tris(diethylamino) analogs show synthetically interesting degrees of syn-selectivity, depending upon the ring size. Optimization (including variation of the N-alkyl group in lactams) remains to be carried out. In all cases the corresponding lithiated species react stereorandomly [67]. In view of these results [8a], it is not surprising that the titanated lactam *292* reacts unselectively with acetaldehyde [79]; $ClTi(NEt_2)_3$ is liable to provide better results.

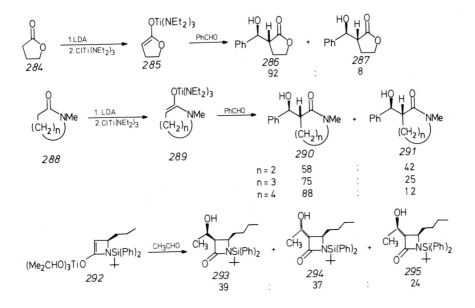

Lithiated *bis*-lactim ethers of the type *296* react with alkyl halides stereoselectively trans to the isopropyl group; hydrolysis affords amino acids in high enantiomeric purity [80]. Unfortunately, this methodology is inefficient if aldehydes are used as the electrophile, since addition is not completely trans to the isopropyl group, and more importantly, simple diastereoselectivity in the aldol-type process is low [80]. This problem was nicely solved by titanating with $ClTi(NMe_2)_3$ (or the N,N-diethyl analog) to form *195A* (or the N-titanated form *195B* as discussed in Section 5.2.3). The reagent reacts with acetaldehyde (and other aldehydes) to form essentially a single diastereomer *297* [81]. Hydrolysis affords enantio- and diastereomerically pure D-threonine (*298*) [81]. This means that simple diastereoselectivity is anti, which is reasonable in terms of chair transition states. In *300* there is severe 1,3-quasi-diaxial interaction between the methyl group and the methoxy group. This is not the case in *299*, which in fact is in line with anti-selectivity.

However, it is not clear why a boat transition state should be of higher energy (they appear to be involved in reactions of Ti-enolates (e.g., *289*) which afford syn-adducts). There may be secondary orbital interactions between the aldehyde and the lactim functionality on the left half of the fairly flat heterocycle *299*. In a boat transition state this is not possible. In any case, *195* is the first optically active organotitanium reagent in which the chiral information is embedded in the organyl moiety (as opposed to chiral ligands). Other examples have been reported since (Section 5.5).

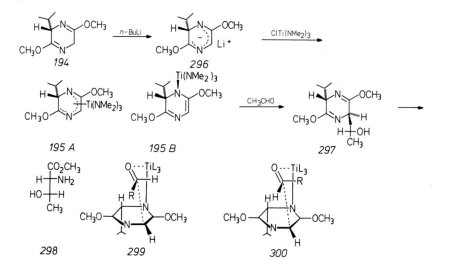

Reagents of the type *195* also add to α-chiral aldehydes with excellent simple diastereoselectivity (anti) and diastereofacial selectivity [58] (Section 5.2.3). This chemistry is an impressive example of how titanation of classical carbanions increases selectivity [8a]. Application in the regio- and stereo-selective addition to α,β-unsaturated aldehydes has also been reported [82].

5.3.3 Addition of Prochiral Allylic Titanium Reagents

The stereoselective addition of substituted allylmetal compounds to carbonyl compounds is of synthetic interest because the adducts can be converted into the corresponding aldols upon ozonolysis [15]. A number of reagents have been employed, notably E- and Z-configurated crotylboron compounds which react stereoselectively with aldehydes to form anti- and syn-adducts, respectively [15, 83]. Nevertheless, other metal systems have been tested, many of them with considerable success [15]. One of the drawbacks of most of these reagents has to to with the fact that addition to ketones fails chemically or occurs non-stereoselectively.

The titanation of crotylmagnesium chloride (which is a mixture of cis/trans *301* and *302*) occurs stereoconvergently to provide essentially a single reagent *303* or its Z-isomer [8a, 20, 23]. Although the vicinal coupling constant of the olefinic protons (12.8 Hz) does not allow for an unambiguous assignment, the assumption of E-configuration seems plausible. Currently, no structural information is available concerning the nondistillable alkoxytitanium compounds *304* and *305*.

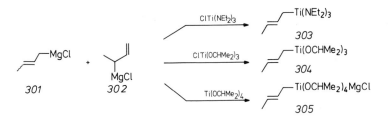

The addition of *303–305* to aldehydes proceeds almost quantitatively and with complete regioselectivity to afford anti- and syn adducts *306* and *307*, respectively [8a, 20, 23]. As shown in Table 4, simple diastereoselectivity varies somewhat according to the type of ligand at titanium, but anti-adducts predominate in all cases. A general conclusion regarding the best ligand system cannot be drawn. In case of aromatic aldehydes such as benzaldehyde, the ate complex *305* is the reagent of choice, but in the aliphatic series the aminotitanium reagent *303* is more selective [8a, 20, 23]. An independent study has shown that crotyltriphenoxytitanium reacts with benzaldehyde to affort an 85:15 anti/syn ratio [84a]. Thus, in this case the more easily accessible ate complex *305* is certainly to be preferred. However, in other cases the triphenoxy ligand system is more selective [84a]. Mechanistically significant is the observation that the nature of the N-alkyl groups of the reagents of the type *303* affects stereoselectivity. For example, tris-(dimethylamino) analog of *303* adds to cyclohexane carboxaldehyde to afford a 66:34 ratio of *306f*/*307f*, whereas *303* leads to an 88:12 ratio (Table 4). Thus, diastereoselectivity decreases in going from *303* to a less bulky reagent. In conclusion, the readily available crotyltitanium reagents *303–305* and the triphenoxy analog may be useful in certain cases, but a number of other crotylmetal reagents show distinctly higher degrees of diastereoselection [15, 83]. The results are in line with a chair transition state (cf. *214*).

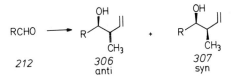

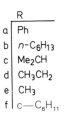

	R
a	Ph
b	$n\text{-}C_6H_{13}$
c	Me_2CH
d	CH_3CH_2
e	CH_3
f	$c\text{-}C_6H_{11}$

Table 4. Diastereoselective Addition of Crotyltitanium Reagents to Aldehydes [8a, 20, 23]

Reagent	Product	anti:syn
303	*306a/307a*	69:31
304	*306a/307a*	80:20
305	*306a/307a*	84:16
303	*306b/307b*	82:*18*
304	*306b/307b*	75:25
305	*306b/307b*	71:29
303	*306c/307c*	85:15
304	*306c/307c*	84:16
305	*306c/307c*	80:20
303	*306d/307d*	85:15
303	*306e/307e*	67:33
303	*306f/307f*	88:12

Real benefits of the above crotyltitanium reagents become apparent in reactions with ketones [8a, 20, 23]. Upon adding *303* to 3-methyl-2-butanone (*308a*), a 97:3 ratio of *309a:310a* was registered ($>95\%$ conversion) [8a]. Several other ketones were also tested with *303*, *304* and *305* [20, 23] (Table 5). In case of acetophenone (308d), the ate complex *305* is best suited, but purely aliphatic ketones require *303* for best results. This is analogous to the trend observed in case of aromatic and aliphatic aldehydes. Crotyltriphenoxytitanium also adds stereoselectively to ketones, but offers no advantages regarding accessibility or degree of simple diastereoselectivity [84b]. It is also noteworthy that the Grignard reagent is non-selective and that allylboron compounds react sluggishly or not at all with ketones [15].

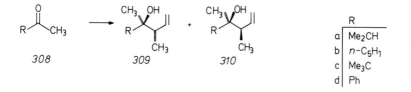

	R
a	Me$_2$CH
b	n-C$_5$H$_1$
c	Me$_3$C
d	Ph

Since the products lack vicinal hydrogens at the two chiral centers, the usual method of anti/syn assignments using ^{1}H-NMR coupling constants cannot be applied here. In case of *309a* and *309d* assignments were made by comparison with authentic samples and/or degradation to known compounds. The tentative assignment of the other adducts is based on the assumption of the same topology in the transition state. A six-membered chair transition state in which the smaller of the two groups flanking the carbonyl function (in the present cases methyl) occupies the pseudo-axial

Table 5. Diastereoselective Addition of
Crotyltitanium Reagents to Ketones
[8a, 20, 23, 84]

Reagent	Product	anti:syn
303	*309a/310a*	97: 3
304	*309a/310a*	88:12
305	*309a/310a*	78:22
303	*309b/310b*	72:28
304	*309b/310b*	67:33
305	*309b/310b*	56:44
303	*309c/310c*	99: 1
304	*309c/310c*	90:10
305	*309c/310c*	92: 8
303	*309d/310d*	85:15
304	*309d/310d*	83:17
305	*309d/310d*	94: 6

position, as in *311*, leads to the observed anti-adducts [20]. The
chair transition state *312* has the bulkier group in the energetically unfavorable axial position and affords the minor syn-adduct. Inspite of the
plausibility of this interpretation, the high degree of simple diastereoselectivity is remarkable. The difference in size between the two groups of the
ketone is not as large as that between the hydrogen and the alkyl (aryl)
group in case of aldehydes, which react less selectively. The explanation
probably has to do with the lower exothermicity of ketone additions, which
means that the transition state comes late (relative to that of aldehyde
addition), resulting in greater compactness and thus greater steric interactions. Unfortunately, it has not been possible to prepare Z-configurated
crotyltitanium reagents.

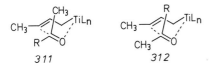

311   *312*

Bis(cyclopentadienyl)titanium(IV) reagents *313* [85] and the related
Ti(III) species *314* [86] add to aldehydes with pronounced anti-selectivity
($>90: <10$ anti/syn ratios). In case of *313*, the chloro derivative does not
behave as selectively as the bromo or iodo analogs, in line with a chair transition state in which 1,3-diaxial interactions are the determining factors.
Reactions with ketones have not been reported.

—TiCp$_2$ a X = Cl   Cp$_2$Ti
313   b X = Br   *314*
          c X = I

165

Recently, reversal of diastereoselectivity in reactions of *313* was achieved by first adding BF_3-etherate to the aldehyde; for example, in case of *212c* the anti:syn ratio changed from 99:1 to 9:91 [87]:

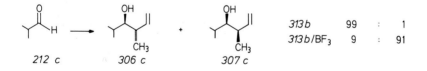

313b	99	:	1
313b/BF₃	9	:	91

212 c *306 c* *307 c*

An open transition state *315* involving the BF_3 complex of the aldehyde is one possible explanation. A related mechanism was first invoked to explain the syn-selective addition of crotylstannanes to BF_3-complexed aldehydes [88]. Structure *315* is in line with the fact that the degree of diastereoselectivity is essentially independent of the nature of the halogen in *313*. However, BF_3 induced ligand exchange processes also have to be considered. More reactive crotyltitanium reagents such as *304* do not show this reversal of stereoselectivity. Since cyclopentadienyl ligands are strongly electron releasing (Chapter 2), Lewis acidity and reactivity of *313* is greatly reduced; thus BF_3 catalysis leading to reversal of diastereoselectivity only operates in case of allyl derivatives which alone react slowly or not at all. A recent study concerning the effect of BF_3 as an additive supports this conclusion [89]. Bulkier Lewis acids (e.g., $TiCl_4$) are not expected to exert the same effect, because the metal ligands should interact sterically with the methyl group of the crotyl reagent (see also *90*). The assumption of anti-complexation of aldehydes (Lewis acid trans to the R-group) is justified by a recent NOE study and X-ray analysis involving the benzaldehyde/BF_3 adduct [90].

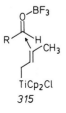

315

Besides the above crotyl compounds, a number of other substituted allylic "carbanions" have been titanated in order to control regio- and stereo-selectivity. An early example concerns titanation of *316* [91]. It was known that *316* reacts with aldehydes and ketones at the γ-position to afford δ-hydroxy vinyl silanes and that addition of MgX_2, ZnX_2 or CdX_2 does not reverse regioselectivity [92]. The problem of clean α-attack was first solved by converting *316* in two steps into a boron reagent (53% yield) and reacting the latter with aldehydes, a sequence which provides anti-adducts *318* preferentially [93]. Shortly thereafter it was found that simply adding $Ti(OCHMe_2)_4$ to *316* has the same effect [91]:

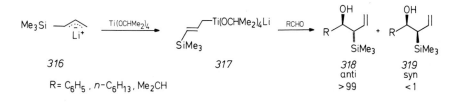

316   317   318   319
                                              anti   syn
R= C₆H₅ , n-C₆H₁₃, Me₂CH   > 99   < 1

Titanation of *316* using ClTi(OCHMe₂)₃ [23], ClTi(NEt₂)₃ [23] or Cp₂TiCl [94] also results in species which react anti-selectively, but the cheap Ti(OCHMe₂)₄ is clearly the titanating agent of choice. Thus, the original idea of generating and using titanium ate complexes [26] is fruitful in case of allylic carbanions, but not so much in the area of enolate chemistry [26, 67]. Reagent *317* also reacts stereoselectively with ketones (e.g., with acetophenone to provide a 91:9 anti/syn product distribution [91]. The adducts are synthetically useful because stereospecific conversion into dienes *322* or *323* is possible using the Peterson olefination under basic or acidic consitions (syn or anti elimination, respectively) [20, 91]. Prolonged reaction times in the addition of *317* lead directly to the corresponding dienes.

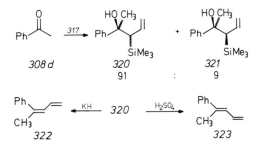

308 d   320   321
          91   :   9

322   323

An interesting application of titanium ate complexes involves *325*, which converts aldehydes and ketones into the anti-adducts stereoselectively, e.g., *326* [95, 96]. Such olefinic β-hydroxy sulfides are useful synthetic intermediates in the synthesis of stereochemically pure alkenyl oxiranes and 2-(arylthio)-1,3-alkadienes.

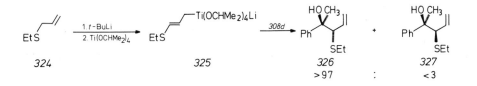

324   325   326   327
                                > 97 : < 3

α-Regioselectivity is also observed in reactions of α- and β-monosubstituted derivatives of *325* (e.g., *329*) [95]. The lithium precursor of *329* does not react regio- or stereoselectively.

5. Stereoselectivity in the Addition of Organotitanium Reagents

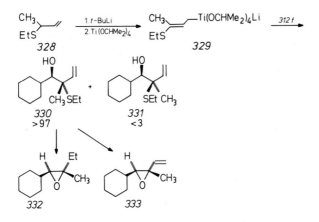

In contrast to the reaction mode of *329*, the opposite regioselectivity is observed if γ-substituted sulfides are used (93–99% γ-selectivity) [95]. In case of *335* [97a], a process analogous to the reactions of *317* [91] occurs. However, the products are not β-hydroxy silanes, but rather the Peterson elimination products *336* themselves. Apparently, *335* adds anti-selectively and the adducts undergo a cis-stereospecific elimination to form *336*. The "discrepancy" may have to do with the different reaction conditions chosen (*317*: −78 °C/1 h [20, 91] vs. *335*: −78 °C/2 h followed by room temperature overnight [97a]).

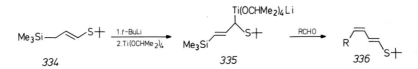

This one pot procedure was elegantly applied to the synthesis of spilanthol *340*, a naturally occurring insecticide from *Spilanthes oleranceae* [97b]. Application in pheromone chemistry also turned out to be successful [97b].

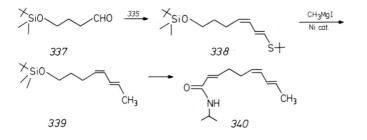

It is also possible to alter the substitution pattern of silylated allylic carbanions and still control regio- and stereoselectivity via titanation. Whereas the oxygen substituted lithium reagent *342* fails to react regioselectively with aldehydes, the ate complex *343* affords isomerically pure dienes *344* (−78 °C, then room temperature (14 h) [98]). Again, this can be rationalized by assuming stereoselective addition. So far the stereoselectivity of this three carbon elongation reaction has not been exploited. However, the regioselectivity characteristic of *343* was used as the controlling element in the total synthesis of the diterpene (±)-aplysin-20 (*348*) [99]. The decisive one-pot sequence *345→346→347* is not only completely regioselective; the quantitative yield also means complete chemoselectivity (Chapter 3).

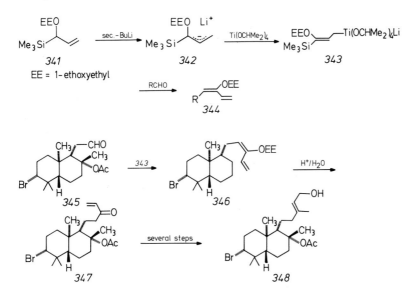

EE = 1-ethoxyethyl

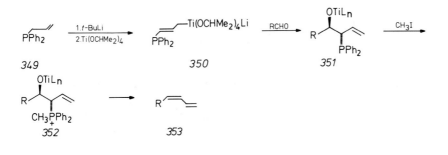

Besides silicon serving as the deoxygenating element, phosphorous can also be employed. Thus, *350* reacts with aldehydes to provide exclusively the anti-adducts *351* [100]. Quaternization at phosphorous generates *352* which fragments directly to the Z-dienes, all in a one-pot procedure. A clear *limitation* of titanium chemistry concerns a related scheme using the analogous diphenylphosphine oxide, which fails to afford appreciable yields of dienes, the reason being unclear [100]. It should be recalled that *353* and the E-isomers are also accessible from the reaction of aldehydes with *317*. In this case the products are the β-hydroxy silanes, which can be transformed either into Z- or E-dienes using the Peterson elimination (see above).

169

5. Stereoselectivity in the Addition of Organotitanium Reagents

Another case of how titanation affects regio- and stereoselectivity concerns the homo-aldol reaction of *354* with aldehydes, which provides almost exclusively the (Z)-anti-adducts, e.g., *355* [101].

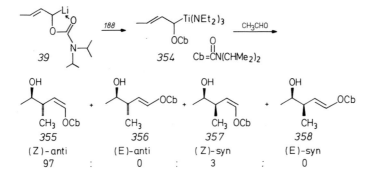

Excellent anti-selectivity also results if *39* is titanated with $ClTi(OCHMe_2)_3$ or $Ti(OCHMe_2)_4$ [25, 101, 102]. The latter two lead to identical results, which may mean that in this system the intermediate ate complex dissociates to the neutral organyltitanium triisopropoxide prior to reaction with aldehydes. In case of additions to ketones, the tris(dialkylamino) ligands are better suited due to the higher regioselectivity [25]. It is, interesting to note that the lithium precursor *39* reacts fairly non-selectively [25]. Whereas metal-metal-exchange using $FB(OCH_3)_2$ does not improve matters, conversion to the aluminum analog by reaction with $ClAl(i-Bu)_2$ results in a product ratio of *355*:*356*:*357*:*358* = 13:78:1:8 [25].

Since the adducts are usually converted into the homo-aldols, the stereochemistry of the double bond is not that important. Titanium is clearly the metal of choice for the selective formation of anti-homoaldol adducts. However, if the Z-configurated analog of *39* is titanated, poor results are obtained, in contrast to Li—Al exchange which leads to preferential formation of syn-adducts [25]. The three-carbon-extension based on the homo-enolate equivalent *354* is synthetically useful, because it provides (inter alia) diastereoselective access to di- or tri-substituted 4-butanolides [102], e.g., *360*. These not only occur in certain natural products, but are also versatile intermediates. Reagents of the type *354* also add stereoselectively to α-chiral aldehydes [25] (Sections 5.2.1 and 5.2.3).

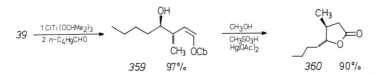

The above methodology for stereoselective homo-aldol additions has been extended to the silylated species *362*, which reacts with acetaldehyde to provide a single diastereomer *363* [103]. Again, the observed anti-selectivity follows the known behaviour of the parent compount *317* [20, 91]. Sub-

170

sequent Peterson elimination under basic or acidic conditions along the lines previously outlined for *317* [91] provides an elegant entry into >99% isomerically pure (1Z, 3Z)- or (1Z, 3E)-dienes *364* and *365*, respectively, which are useful in Diels-Alder reactions [103].

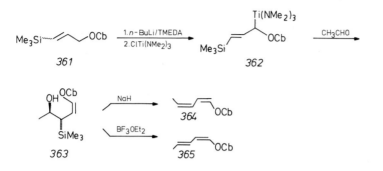

A large number of other heteroatom-substituted allylic carbanions have been prepared; again, regio- and stereoselectivity in reactions with aldehydes are not uniformly acceptable [104]. It is likely that carbanion-selectivity can be controlled via titanation in many of these cases, particularly in view of the fact that variation of ligands at titanium is a simple matter [8a] (Chapter 1).

An example concerns *367*, in which variation of ligands remains to be studied [31].

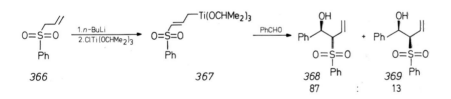

In summary, titanation of prochiral allylic anions using Ti(OCHMe$_2$)$_4$, ClTi(OCHMe$_2$)$_3$ or ClTi(TiNR$_2$)$_3$ constitutes a simple means to control regio- and stereoselectivity (anti-adducts). The latter can be explained by assuming a chair transition state of the type *214* (M = TiLn) which provides the anti-adducts. So far, Cp ligands at titanium have not proved to be as versatile, although they are useful in certain cases. Titanation of optically active amino-substituted allylic carbanions ensures complete control of regio- and enantioselectivity [105] (Section 5.5).

5.3.4 Addition of Prochiral Propargyl-Titanium Reagents

Propargyl anions are of potential synthetic interest in carbonyl chemistry as carbon-chain extending synthons, because the adducts contain functionality which can be manipulated in useful ways [106]. However, application in organic synthesis has been limited because of the difficulties in controlling regio- and stereoselectivity. Due to the propargylic/allenic anion equilibrium [107], addition to aldehydes may form either acetylenic or allenic alcohols. Besides the problem of regiocontrol, each of the products may consist of two stereomers (anti or syn). As expected, the nature of the metal is of prime importance.

In a recent definitive study, the lithium in the propargylic species *371* was exchanged for other metals and the regioselectivity in the addition to cyclohexanecarbaldehyde studied [108]. Table 6 clearly shows that titanation using $Ti(OCHMe_2)_4$ in THF is the method of choice for selectively obtaining the α-allenic alcohol *374*. This is best explained by assuming the intermediacy of a propargyltitanium species *372*; however, it was not possible to gain NMR spectroscopic evidence [108].

$$CH_3C\equiv CCH_3 \xrightarrow{t\text{-}BuLi} (CH_3C\equiv C-CH_2)Li \xrightarrow{Ti(OCHMe_2)_4} CH_3C\equiv CCH_2Ti(OCHMe_2)_4Li$$
$$\quad\quad 370 \quad\quad\quad\quad\quad\quad 371 \quad\quad\quad\quad\quad\quad\quad\quad\quad 372$$

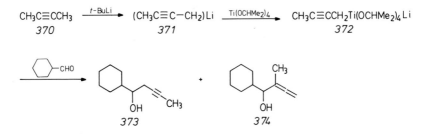

373   *374*

Following these exploratory experiments, substituted acetylenes *375* were employed [108]. It turned out that for $R^2 = H$, only α-allenic alcohols *380* result. However, if $R^2 = CH_3$, the opposite regioselectivity pertains, affording diastereomeric β-acetylenic alcohols *378/379*. This can be explained on the basis of steric repulsion between R^2 and the titanium moiety, making *376* energetically unfavorable; *377* then reacts with "allenic inversion".

Table 6. Regioselectivity of the Addition of *371* to Cyclohexanecarbaldehyde as a Function of Additives [108]

Additive	Solvent	*373:374*
—	Ether	34:66
—	THF	58:42
$Ti(OCHMe_2)_4$	Ether	3:97
$Ti(OCHMe_2)_4$	THF	1:99
$B(OCHMe_2)_3$	Ether	69:31
$B(OCHMe_2)_3$	THF	75:25
Et_2AlCl	THF	10:90
MgI_2	THF	19:81
$ZnBr_2$	Ether	9:91
$ZnBr_2$	THF	17:83
$SnCl_2$	THF	7:93

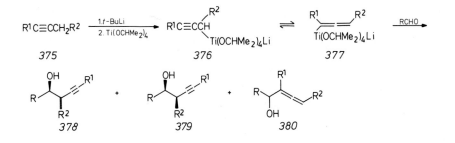

375   376   377

378   379   380

Reagents *377* having various R^1 and R^2 groups add to aldehydes anti-selectively (better than 90:10 product ratios of *378/379* with no formation of *380*). R^2 can be methyl, OTHP or $OCMe_2OMe$ [108]. An unexplained exception is the reaction with benzaldehyde, which delivers a greater proportion of the syn-adduct. Otherwise, the results are best explained by the two cyclic transition states *381* and *382*, the latter being of higher energy due to steric repulsion between R^2 and R. It is noteworthy that selectivity of the corresponding Li, Mg and Zn reagents is considerably lower [108a].

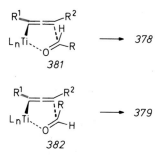

381   ⟶   378

382   ⟶   379

This methodology has been applied to the stereocontrolled synthesis of (±)-asperlin (*385*) and related compounds [109].

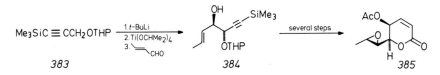

383   384   385

Similarly, the bis (silyl) derivative *386* reacts with aldehydes in a one-pot procedure to provide Z-enynes *387* exclusively [108a]. Apparently, the reagent again prefers the allenic structure and reacts anti-selectively, the adduct undergoing a *cis*-stereospecific Peterson elimination under the reaction conditions. Related additions to imines are also synthetically useful [110a].

386   387

173

5. Stereoselectivity in the Addition of Organotitanium Reagents

Besides acetylenes, allenes can also serve as starting materials. A simple sequence based on the metallation of *388* is an example [110b]. The relatively low diastereoselectivity may yet be improved by using other ligands at titanium.

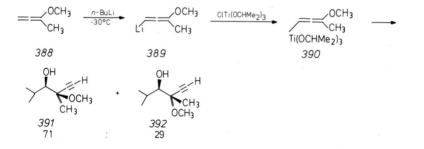

| 388 | 389 | 390 |

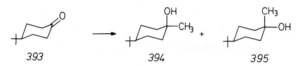

| 391 | 392 |
| 71 | 29 |

Finally, lithiated α-silyl acetonitrile (which is structurally related to acetylenes) shows moderate diastereoselectivity in reactions with aldehydes. Titanation using Ti(OCHMe$_2$)$_4$ increases stereoselectivity to >90%; however, an analogous sequence employing B(OCHMe$_2$)$_3$ is even more selective [108a].

5.4 The Problem of Equatorial vs. Axial Addition to Cyclic Ketones

The question of axial versus equatorial addition of organometallics to cyclic ketones has been addressed on numerous occasions [111]. In case of six-membered rings, conformationally locked 4-*tert*-butylcyclohexanone (*393*) has often served as a convenient model compound. The traditional explanation for the stereochemical outcome involves two opposing factors [111]:
1) Non-bonded interaction between the incoming nucleophile and the 3,5-axial hydrogen atoms in case of axial attack; and
2) torsional strain arising from interaction between the incoming nucleophile and the 2,6-hydrogen atoms.

In most cases the former is more severe, so that equatorial attack with formation of the axial alcohol predominates. The results of some methyl-metal additions are summarized in Table 7.

| 393 | 394 | 395 |

The highest degree of diastereoselectivity results in case of CH$_3$Ti(OCHMe$_2$)$_3$ (*12*) in hexane at −15 °C → +22 °C which leads to a 94:6 ratio of *394:395* (entry 7). Two equivalents of CH$_3$Li in the

174

Table 7. Addition of Methylmetal Reagents to 4-*tert*-Butylcyclohexanone[a]

Entry	Reagent	Solvent	Temp. (°C)	*394:395*	Ref.
1	$(CH_3)_2TiCl_2$	CH_2Cl_2	−78	82:18	18
2	$CH_3Li/TiCl_4$	ether	−10	70:30	18b
3	$CH_3Ti(OCHMe_2)_3$	CH_2Cl_2	+22	82:18	8a
4	$CH_3Ti(OCHMe_2)_3$	ether	+22	86:14	8a
5	$CH_3Ti(OCHMe_2)_3$	ether	0	89:11	8a
6	$CH_3Ti(OCHMe_2)_3$	CH_3CN	?	86:14	8b
7	$CH_3Ti(OCHMe_2)_3$	*n*-hexane	−15 → +22	94: 6	8a, 20
8	$CH_3Li/Ti(OCHMe_2)_4$	ether	−15 → +22	—[b]	24
9	$(CH_3)_4Ti$	ether	−50	38:62	8a, 24
10	$CH_3MgI/Ti(OCHMe_2)_4$	ether	−40	33:67	8a, 24
11	CH_3MgI	ether	0	63:38	112
12	CH_3Li	ether	5	65:35	113
13	$2 CH_3Li/LiClO_4$	ether	−78	92: 8	114
14	$2 CH_3Li/3 (CH_3)_2CuLi$	ether	−70	94: 6	115
15	$(CH_3)_3Al$ (1 part)	benzene	+22	76:24	116
16	$(CH_3)_3Al$ (3 parts)	benzene	+22	12:88	116
17	$CH_3CdCl + MgX_2$	ether	0	38:62	112
18	$(CH_3)_2Zn + MgX_2$	ether	0	38:62	112
19	$CH_3Zr(OBu)_3 + LiCl$	ether	+22	80:20	117

[a] Conversion >85% in all cases involving titanium with the exception of entry 8.
[b] This ate complex does not undergo addition to *373*; instead, condensation products are formed.

presence of $LiClO_4$ (entry 13) or a mixture of CH_3Li and $(CH_3)_2CuLi$ (entry 14) are just as selective, but require an excess of active methyl groups. Thus, $CH_3Ti(OCHMe_2)_3$ is currently the reagent of choice. Steric interaction of this bulky reagent with the 3,5-hydrogen atom outweighs torsional effects. A striking difference was observed upon using tetramethyltitanium (entry 9) or $CH_3MgI/Ti(OCHMe_2)_4$ (entry 10), the degree of attack at the sterically more hindered side being significant. This is reminiscent of Ashby's observation that a threefold excess of $(CH_3)_3Al$ leads to reversal of diastereoselectivity (entry 16). He introduced the concept of compression, i.e., "the effective bulk of the carbonyl group increases to such an extent by complexation with an organoaluminum compound that severe interaction with groups on adjacent carbons can occur in the transition state" [111]. This may occur whenever the reagent complexes with the carbonyl group followed by the addition of a second molecule of reagent. A similar phenomenon may be operating to some extent in the titanium cases (entries 9 and 10). Nevertheless, finding an economical method to achieve clean axial attack remains a challenge.

Methyltitanium reagents have been used in steroid chemistry. $CH_3Ti(OCHMe_2)_3$ (*12*) adds to 3-cholestanone selectively from the equatorial direction to provide an 87:13 mixture of the corresponding α- and β-alcohol, respectively [20]. Androstan-3,17-dione (*396*) reacts chemo- and stereo-selectively [18b]:

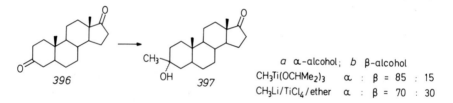

			a α-alcohol; *b* β-alcohol
CH₃Ti(OCHMe₂)₃	α	:	β = 85 : 15
CH₃Li/TiCl₄/ether	α	:	β = 70 : 30

Allylmetal reagents sometimes add to *393* preferentially from the axial direction, but the degree of stereoselection is poor [111]. In contrast, allyltris(diethylamino)titanium (*36*) affords a product ratio of *398:399* = 20:80 [20]. This is best explained by assuming that non-bonded interaction with the 3,5-hydrogen atoms is less severe than torsional strain between the 2,6-hydrogen atoms and the bulky amino ligands (*400* vs. *401*). Thus, the use of bulky titanium reagents allows for reversal of diastereoselectivity relative to the reaction of the "slender" diallylzinc (*398:399* = 84:16) [119] or the BF₃-mediated addition of CH₂=CHCH₂Sn(n-C₄H₉)₃ [120].

293 →

	398		*399*	Ref.
CH₂=CHCH₂MgCl	45	:	55	118
(CH₂=CHCH₂)₂Mg	44	:	56	119
CH₂=CHCH₂Ti(OCHMe₂)₃	43	:	57	20
CH₂=CHCH₂Ti(NEt₂)₃	20	:	80	21
(CH₂=CHCH₂)₂Zn	84	:	16	119
CH₂=CHCH₂Sn(n-C₄H₉)₃/BF₃	92	:	8	120

400   *401*

A limitation of titanium chemistry concerns the failure of *n*-alkyl reagents to add cleanly to ketones (Chapter 3). Also, the titanation of phenyllithium does not increase stereoselectivity in the addition to *393* [8b]. The vast majority of "resonance stabilized" titanium reagents currently known have not yet been tested in reactions with cyclic ketones such as *393*.

The pronounced tendency of organometallics to attack 2-methyl-cyclohexanone (*402*) from the equatorial direction has been ascribed to increased steric interaction of the pseudo-axial hydrogen of the methyl group with axially incoming nucleophiles [111]. Thus, CH_3Li, CH_3MgCl and PhMgBr deliver *403: 404* ratios of 88:12, 88:12 and 91:9, respectively [111], and titanium reagents are even more selective [8a, 20, 21]. Here again, 2 $CH_3Li/LiClO_4$ and 2 $PhLi/LiClO_4$ are also >94% stereoselective, but require an excess of active methyl or phenyl groups [114, 115]. As in case of *393*, the stereoselective introduction of allyl groups is more difficult. Allyltriisopropoxytitanium (*33*) affords a 76:24 product ratio of *403:404* (R = allyl), respectively, which is identical to the performance of diallylmagnesium [119]. Salt-free diallylzinc, prepared from $(CH_3)_2Zn$ and triallylboron shows greater diastereoselectivity (89:11 ratio) [119].

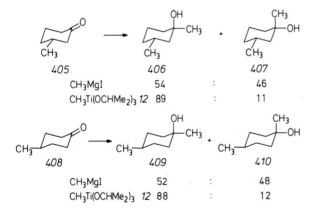

$CH_3Ti(OCHMe_2)_3$	12	96	:	4	
$PhTi(OCHMe_2)_3$	38	> 99	:	< 1	
$CH_2=CHCH_2Ti(OCHMe_2)_3$	33	76	:	24	

In case of conformationally labile cyclohexanones *405* and *408*, classical reagents often react stereorandomly [111, 121]. The diastereoselectivities achieved by titanium analogs are therefore remarkable [8a, 20].

	405	406		407
CH_3MgI		54	:	46
$CH_3Ti(OCHMe_2)_3$	12	89	:	11

	408	409		410
CH_3MgI		52	:	48
$CH_3Ti(OCHMe_2)_3$	12	88	:	12

In contrast to cyclohexanone derivatives, cyclopentanones have not been studied systematically in reactions with titanium reagents [121]. In case of sterically hindered 2-phenyl-2-methylcyclopentanone (*1*), $(CH_3)_2TiCl_2$ adds cis to the phenyl group, possibly due to prior complexation (Sections 5.1). However, this does not extend to 2-phenylcyclopentanone (*411*). In this case, $(CH_3)_4Ti$ leads to the largest proportion of *412*; it is currently difficult to decide whether this is due to some sort of electronic interaction between the phenyl ring and the Lewis acidic titanium reagent.

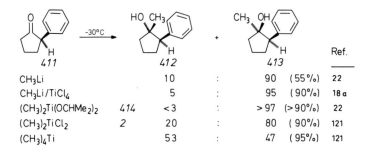

	412		413		Ref.
CH₃Li	10	:	90	(55%)	22
CH₃Li/TiCl₄	5	:	95	(90%)	18 a
(CH₃)₂Ti(OCHMe₂)₂ 414	<3	:	>97	(>90%)	22
(CH₃)₂TiCl₂ 2	20	:	80	(90%)	121
(CH₃)₄Ti	53	:	47	(95%)	121

5.5 Enantioselective Additions

Various attempts have been made to perform enantioselective Grignard-type additions of RLi and RMgX by prior complexation with chiral amines or ethers [122]. High ee-values (enantiomeric excess) are rare, probably due to the fact that the chiral auxiliary is not bound intimately to the organo-metallic reagent, and that more than one species can be involved. Thus, it seems that monomeric σ-bonded titanium compounds of well defined geometry should be ideal reagents. In principle, there are four ways to approach this problem [121]:

1) Using compounds having chiral alkoxy or amino ligands.
2) Using compounds containing a center of chirality at titanium.
3) Using compounds incorporating both structural features 1) and 2).
4) Using reagents RTiX₃ in which the chiral information is contained in the organyl moiety R.

Strategies 1–3) are the most general, since in principle any carbanion could be transformed into an optically active titanium reagent. So far, the success has been limited, although recent developments may lead to better results. By nature, strategy 4) represents special cases, since the products of carbonyl addition always contain a certain type of functionality originating from the chiral carbanionic precursor. Several impressive examples have in fact been reported, and more are likely to follow.

5.5.1 Reagents with Chirally Modified Ligands at Titanium

A mild and salt-free method for preparing methyl- or phenyltitanium reagents having chiral alkoxy (or amino) groups is to treat compounds of the type R₂Ti(OR′)₂ with one equivalent of an optically active alcohol (or amine). For example, menthol (415) protonates 414 to form methane and 416 in quantitative yield [22, 121]. Reaction with benzaldehyde produces 417 cleanly, but enantioselectivity is poor (ee = 13.5%); S-configuration predominates [22, 121]:

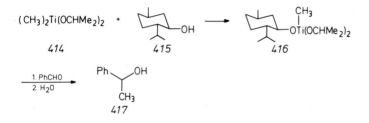

414   415   416

417

A second method of synthesizing chiral titanium compounds involves substitution of chlorine by RLi in titanating agents already containing optically active alkoxy groups, e.g., *418* → *419* [121]: In this case the reaction with benzaldehyde leads to an ee-value of 18% (S-enantiomer in excess) [121]:

$$\text{ClTi} \left(\text{O} \underset{418}{} \right)_3 \xrightarrow{\text{CH}_3\text{Li}} \text{CH}_3\text{Ti} \left(\text{O} \underset{419}{} \right)_3$$

Similarly, *421* leads to an ee-value of 8% in the reaction with benzaldehyde [123].

$$\text{ClTi} \left(\text{O} \underset{420}{} \right)_3 \xrightarrow{\text{CH}_3\text{Li}} \text{CH}_3\text{Ti} \left(\text{O} \underset{421}{} \right)_3$$

Better results are obtained by using S-(—)-β-binaphthol *422* as the ligand system [8a, 121, 124]. In case of phenyl transfer *424* → *425*, the ee-value is 88%, an unprecendented result in titanium chemistry [124].

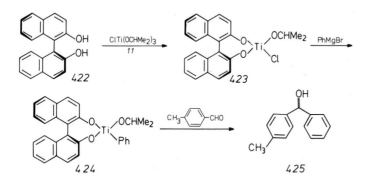

422   423

424   425

Unfortunately, such high ee-values are the exception. Related compounds *426* add to aldehydes with varying degrees of enantioselectivity (ee = 10–76%) [121, 125]. Part of the problem has to do with the fact that some of the reagents *426* are either aggregates or more likely oligomers (or both). The ^{1}H- and ^{13}C-NMR spectra show that in several cases a number of different species are actually involved [121].

5. Stereoselectivity in the Addition of Organotitanium Reagents

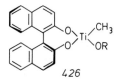

426

R = Et, CHMe₂, CMe₃

Since oligomerization (as well as aggregation) is facilitated if Lewis acidity of titanium is pronounced (among other factors), attempts were undertaken to prepare cyclic titanium compounds having Cp groups as ligands [126]. These groups are strongly electron releasing, reducing Lewis acidity (Chapter 2). At the same time they shield sterically, which also makes intermolecular Ti-O interactions less likely. To this end, *422* was transformed into *427* (80% yield; melting point 274 °C with dec.), the first well characterized titanium derivative in the binaphthol series. Besides a correct elemental analysis, the ^{1}H- and ^{13}C-NMR spectra clearly show that the compound is monomeric [126]. The $[\alpha]_D^{22}$ value is $+1505$ °C (c = 1.11, CH₂Cl₂).
The CD spectrum is shown in Fig. 4.

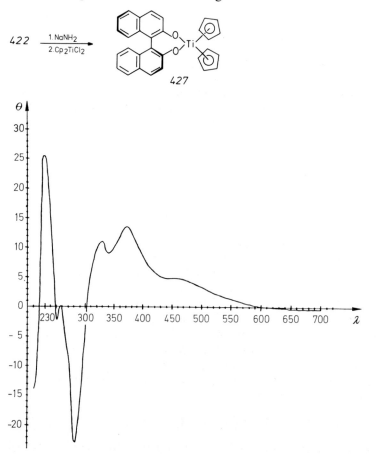

Fig. 4. CD Spectrum of *427* [126].

180

Unfortunately, related attempts to synthesize a chiral titanating agent *429* for carbanions failed, mixtures of various (oligomeric?) products being obtained [126]. In contrast, chlorination of the Ti(III) species *430* with CCl₄ afforded a monomeric species *429*, but is was not possible to isolate the compound in pure form. Oligomerization sets in within a week. In situ allylation followed by addition of benzaldehyde and aqueous workup led to predominant formation of S- *432* (ee = 30%) [126]. This disappointing result may be due to the presence of species other than the desired *431*.

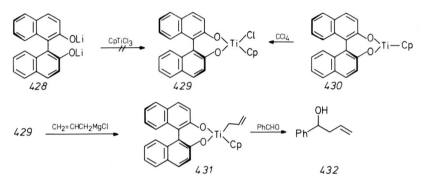

It seems that titanium(IV) has no great propriety to be part of such seven-membered rings. However, this cannot be generalized on the basis of current information. Even greater difficulties were encountered in attempts to prepare five-membered cyclic titanium compounds of the type *434*. Oligomerization and other reactions predominate, as shown by careful NMR experiments [31, 126].

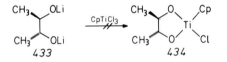

In many of these undefined systems, additions of CH_3Li or $CH_2 = CHCH_2MgCl$ leads to species which react enantioselectively with aldehydes (ee = 10–50%), but the results are of no preparative or mechanistic value [31]. All of these observations are related to some of the erratic results obtained in case of *426* [121, 125]. Thus NMR characterization should always be carried out. In summary, a few specific cases of acceptable ee-values have been reported, but the problem of enantioselective addition of titanium reagents remains unsolved. It is interesting to note that six-membered ring systems have not been reported in such reactions, inspite of the fact that cyclic titanium compounds derived from 1,3-diols are known (Chapter 2).

5.5.2 Reagents with the Center of Chirality at Titanium

A different approach to enantioselective Grignard- (or aldol) type of addition in the area of titanium chemistry is based on reagents having a center of chirality at the metal. This places the chiral information in closest vicinity to the carbonyl function during the process of addition [8a]:

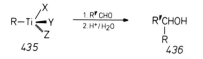

The basic problem is to find ligands X, Y and Z that impart configurational stability upon the tetrahedral titanium, but also ensure enough reactivity in reaction with aldehydes. Unfortunately, these two factors oppose each other [121, 127]. For example, racemic *437* (here and in the following examples only one enantiomeric form is arbitrarily shown) adds to benzaldehyde in ether at room temperature within 30 minutes to from ~95% of the addition product *417* (following aqueous workup) [127]. However, its ^{1}H-NMR spectrum displays only one doublet for the formally diastereotopic methyl groups of the isopropoxy ligand. Cooling to −30 °C simply leads to line broadening of all signals, which points to aggregation phenomena (Chapter 2). Although the appearance of a single clean doublet at room temperature may be accidental, other similar Ti(IV) compounds display the same effect. It is likely to be due to bimolecular ligand exchange reactions, in which two species interact via Ti—O—Ti bridging. As noted in Chapter 2, certain dimeric species have been shown to have such bridging in which titanium is penta-coordinated. Even if chiral compounds of the type *437* or *438* are not dimeric (which has not been looked into) [121], short-lived intermediates involving similar bonding are likely to occur which lead to configurational instability. An even more serious problem is that the dimers may lead to ligand exchange so as to generate new compounds which are no longer chiral. Upon prolonged standing compounds such as *437* do in fact begin to decompose, possibly via dismutation. Thus, ligands have to be chosen which reduce Lewis acidity (in this case oxophilicity) and/or shield sterically so as to prevent Ti—O—Ti bridging. Since it is known that pentahapto cyclopentadienyl groups exert such effects, compounds of the type *441* were synthesized [121].

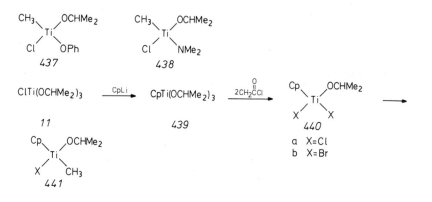

The compounds *441a–b* add smoothly to aldehydes. Also, the configurational stability is considerably higher than that of chiral titanium compounds lacking Cp ligands [121, 127]. The ^{1}H-NMR spectrum in the temperature range of 0 °C to −30 °C shows two resolved doublets for the diastereotopic methyl groups of *441a* (Fig. 5). The other signals are also sharp, which speaks for a single monomeric species. However, at higher temperatures dynamic effects are observed, i.e., the two doublets cleanly coalesce to one doublet. $\Delta G^{\pm}$ for enantiomerization amounts to 19.2 kcal/mol (80.3 kJ/mol) [121, 127]. The bromide *442b* behaves similarly. Thus, separation into antipodes is not feasible.

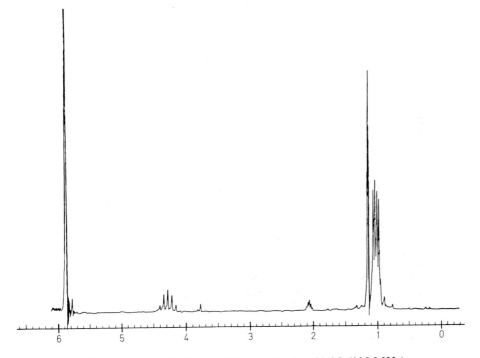

Fig. 5. ^{1}H-NMR Spectrum of *441a* in Toluene-D$_8$ at −30 °C (100 MHz)

Upon replacing the chlorine in *441a* by (trimethylsilyl)methyl, a compound *443* was obtained whose ^{1}H-NMR spectrum shows two doublets for the diastereotopic methyl groups in the range of −30 °C to +100 °C [121]. At the upper temperatures slow decomposition begins. At +40 °C *443* adds to benzaldehyde to form *417* (60% after 12 h) following aqueous workup [121]. The diastereomer ratio of the titanium containing adducts prior the hydrolysis remains to be determined. These results show that compounds of the type *443* may be separable into antipodes. However, initial efforts involving the menthoxy analog of *443* were not rewarding because the diastereomers could not be crystallized or separated by other means [31].

183

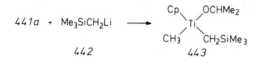

441a + Me₃SiCH₂Li ⟶ 443

442

Parallel to these efforts, titanium compounds containing two Cp groups were tested [121]. Although (racemic) *445* appears to be configurationally stable as ascertained by high temperature NMR measurements, it fails to add to aldehydes (22 °C/7 days). The electron-releasing effect of two Cp groups (Chapter 2) is so pronounced that carbonylophilicity is drastically reduced.

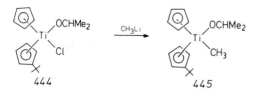

444   445

A way out of this dilemma is to use more reactive bis(cyclopentadienyl)-titanium compounds. Since allyltitanium reagents are considerably more reactive than the methyl or phenyl analogs [8a] (Chapter 4), it seemed logical to test such compounds. Previously, substituted derivatives *446* had been prepared and isolated in optically active (+) and (−) forms [128]. The addition of allylmagnesium chloride to racemice *446* afforded *447*, a sensitive compound which was reacted in situ with various aldehydes in hope of obtaining diastereomeric adducts *448*. Indeed, conversion at −78 °C to room temperature is ≥85% [8a, 127]. Compounds *448* are not air-sensitive and can be isolated and characterized. In case of *448a* the ¹H-NMR spectrum has been published [127]. Hydrolysis using saturated aqueous NH₄F solutions liberates *449*. The diastereomer ratios of *448* are recorded in Table 8.

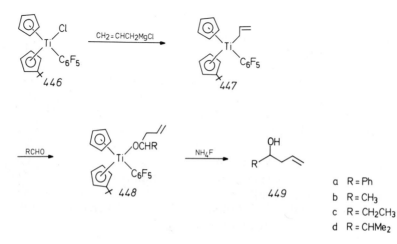

a   R = Ph
b   R = CH₃
c   R = CH₂CH₃
d   R = CHMe₂

Table 8. Diastereoselective Reaction of *447* with Aldehydes RCHO to Form *448* [127]

Product	Diastereomer-Ratio	de (%)
448a	60:40	20
448b	70:30	40
448c	64:36	28
448d	56:44	12

The low degree of stereoselection shows that the two Cp ligands are not sufficiently different in steric or electronic nature. In principle, this can be controlled by choosing the proper substituents at one of the Cp groups (cf. Chapter 2). Nevertheless, irrespective of this controlling element, another problem presents itself. The reaction of allylmagnesium chloride with the titanating agent must be stereospecific either with inversion or retention of configuration at titanium. To check this, optically active ($+$)-*446* [128] was treated with allylmagnesium chloride, benzaldehyde added and the product cleaved with NH$_4$F in a one-pot sequence [8a, 127]. The final product *432* turned out to have the S-configuration, the ee-value being 11%. This is the first example of an enantioselective Grignard-type addition in which the metal is the chiral center [8a]. Theoretically, the de-value in the racemic series (Table 8) is the maximum ee-value. Since the former is 20%, loss of chiral information is occurring. Partial racemization during the substitution process using allylmagnesium chloride is plausible [127].

($+$)-*446* → [1. CH$_2$=CHCH$_2$MgCl 2. PhCHO 3. NH$_4$F] → *432* (Ph, OH, allyl product)

These experiments show that care must be taken so that the carbon nucleophile transfers onto titanium stereospecifically. If this is not possible, one (or more) of the ligands should be optically active. This means that the diastereomers, each having chirality centers at titanium and in the ligand, would have to be separated prior to reaction with aldehydes. Since this may pose serious problems (cf. menthoxy analogs of *443*), a new strategy was recently considered: Use of optically active bidentate ligand systems which result in a single diastereomer upon attachment of titanium [31]. This is theoretically possible in cyclic or bicyclic system such as *450*, which are likely to be formed stereoselectively because the bulky Cp has room only in the position trans to the R^1/R^2 groups. There is also no need to worry about configurational change at titanium during the introduction of R groups, provided they are not bulkier than Cp ligands.

450   451

The above strategy was first tested in a simple, but not optimum model system [129]. The optically active diol *452* was cleanly converted into *453*. The other diastereomer in which the Cp group is situated endo to the camphor skeleton is not formed; NOE experiments clearly prove that the stereochemistry of *453* is as shown [129]. In this molecule titanium represents a center of chirality. Also, unlike compounds derived from simple diols (e.g., *434*), the five-membered cyclic monomeric titanium species is stable. Oligomerization probably does not occur due to steric reasons. Furthermore, reaction with CH_3Li occurs with complete retention of configuration. Again, the configuration at titanium in *454* was proven by NOE experiments [129]. Unfortunately, *454* reacts sluggishly with aldehydes, delivering adducts with low stereoselectivity.

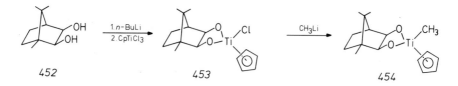

452   453   454

More reactive carbon nucleophiles such as Li-enolates were also treated with *453* and the resulting chiral titanium reagents *456* reacted with aldehydes and ketones [129]. Following aqueous workup, the aldol adducts *457* turned out to be optically active, the ee-values ranging between 8 and 27% (Table 9) [129]. In all cases the preferred direction of attack is Si (the "exception" noted in Table 9 simply involves a switch in priority within the Cahn-Ingold-Prelog-nomenclature).

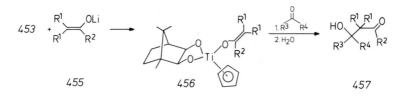

455   456   457

Obviously, the results are of no synthetic value; other methods of enantioselective aldol additions lead to ee-values of $>90\%$ [130]. However, the chiral titanium reagents *456* do represent progress, because they show that the strategy outlined by *450* $\rightarrow$ *451* is feasible. It is also remarkable that ring-opening and oligomerization does not occur. The reason for the (anticipated) low ee-values has to do with the fact that the two alkoxy substituents are too similar in nature. Therefore, efforts directed towards the synthesis of such

Table 9. Enantioselective Aldol Additions Using Chiral Titanium Enolates *456* [129]

R¹	R²	R³	R⁴	ee (%)	Predominant enantiomer	Preferred direction of attack
H	OCH_3	Ph	H	27	S	Si
H	OEt	Ph	H	24	S	Si
H	OtBu	Ph	H	7	S	Si
CH_3	OCH_3	Ph	H	23	R	Si
H	OCH_3	$n\text{-}C_7H_{15}$	H	28	R	Si
H	OCH_3	Ph	CF_3	11	S	Re
H	OCH_3	$c\text{-}C_6H_{11}$	CH_3	2	S	Si

compounds as *458* and *459* are now in progress [129]. Here again, the bulky Cp groups should occupy the sterically less hindered position. However, *459* is likely to polymerize.

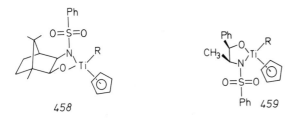

458   *459*

5.5.3 Titanation of Carbanions which Contain a Chiral Auxiliary

Although the above strategies represent the most general case (*460* → *461*), titanation of carbanions containing chiral information themselves may also be of synthetic interest (*462* → *463*). Such manipulation is meaningful only in cases in which the precursor *462* itself reacts non-selectively with aldehydes.

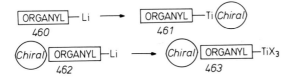

The latter case means that the product of aldehyde or ketone addition carries the chiral auxiliary with it. It may be cleaved in a further step, thereby affording a product containing an additional functionality of specified nature. The first such example (which also involves simple diastereoselectivity) pertains to the addition of titanated bis-lactim ethers (*195*) as discussed in Section 5.3.2. Transfer of chiral information onto the reacting aldehyde is complete [80–82]. Another recent example (devoid of simple diastereoselectivity) is shown below [105]. Titanation of the amino-substituted allyllithium reagent *465* results in *466* which reacts regio- and stereo-selectively with aldehydes and ketones to provide *467* (Table 10). This sequence is the first case of an enantioselective homo-aldol addition. The enamines *467* (which are pure after one crystallization) can be hydrolyzed to the corresponding aldehydes, which form the acetals *469* in the presence of methanol. Oxidation affords enantiomerically pure lactones *470*. Some of these compounds are pheromones, others fragrants [105]. They can also be used for further synthetic transformations. This is another impressive example of how titanation of a carbanion increases selectivity [8a] (*465* and the magnesium counterpart fail to show stereoselectivity) [105]. A cyclic six-membered transition state has been postulated [105].

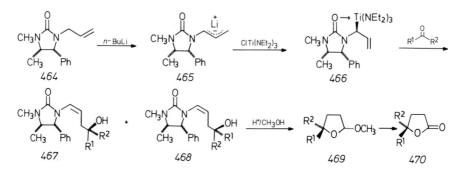

Table 10. Stereoselective Homo-Aldol Addition of *466* to Carbonyl Compounds [105]

R^1	R^2	Yield (%)	*467*:*468*
n-C$_8$H$_{17}$	H	96	94:6
Et	H	94	96:4
CHMe$_2$	H	95	96:4
CH$_3$	CHMe$_2$	93	98:2

Finally, a variation of the strategy outlined by *462* → *463* is possible in which the chiral information is not linked to the carbanionic species via a heteroatom (nitrogen in case of *296* and *465*), but in which the skeleton of the carbanion itself is chiral. For example, certain terpenes have been lithiated and subsequently titanated to produce special reagents which add stereoselectively to aldehydes in a homo-aldol-like fashion [25].

References

1. a) Martens, J.: Top. Curr. Chem. *125*, 165 (1984); b) Morrison, J. D. (editor): "Asymmetric Synthesis", Volumes 1–5, Academic Press, New York, 1983–85; c) Bartlett, P. A.: Tetrahedron *36*, 3 (1980).
2. First communicated in lectures presented by M. T. Reetz at the Universität München (November 10, 1979), Bayer AG, Krefeld (November 15, 1979) and Max-Planck-Institut für Kohleforschung, Mülheim (December 10, 1979).
3. Reetz, M. T.: Nachr. Chem. Techn. Lab. *29*, 165 (1981).
4. Morrison, J. D., Mosher, H. S.: "Asymmetric Organic Reactions", Prentice-Hall, Englewood Cliffs, New Jersey, 1971.
5. Erskine, G. J., Hurst, G. J. B., Weinberg, E. L., Hunter, B. K., McCowan, J. D.: J. Organomet. Chem. *267*, 265 (1984).
6. Dijkgraaf, C.: J. Phys. Chem. *69*, 660 (1965).
7. a) Davies, G. R., Jarvis, J. A. J., Kilbourn, B. T.: J. Chem. Soc., Chem. Commun. *1971*, 1511; b) Bassi, I. W., Allegra, G., Scordamaglia, R., Chioccola, G.: J. Am. Chem. Soc. *93*, 3787 (1971).
8. a) Reetz, M. T.: Top. Curr. Chem. *106*, 1 (1982); additional reviews; b) Bottrill, M., Gavens, P. D., Kelland, J. W., McMeeking, J.: in "Comprehensive Organometallic Chemistry", Wilkinson, G., Stone, F. G. A., Abel, E. W. (editors), Pergamon Press, Oxford, Chapter 22, 1982; c) Weidmann, B., Seebach, D.: Angew. Chem. *95*, 12 (1983); Angew. Chem., Int. Ed. Engl. *22*, 31 (1983); d) Reetz, M. T.: Pure Appl. Chem. *57*, 1781 (1985).
9. Cram, D. J., Abd Elhafez, F. A.: J. Am. Chem. Soc. *74*, 5828 (1952).
10. Cherest, M., Felkin. H.. Prudent. N.: Tetrahedron Lett. *18*, 2199 (1968).
11. Anh, N. T.: Top. Curr. Chem. *88*, 145 (1980).
12. Cornforth, J. W., Cornforth, R. H., Mathew, K. K.: J. Chem. Soc. *1959*, 112.
13. a) Cram, D. J., Kopecky, K. R.: J. Am. Chem. Soc. *81*, 2748 (1959); b) Eliel, E. L.: in ref. [1 b], Vol. 2, 1983.
14. a) Evans, D. A., Nelson, J. V., Taber, T. R.: Top. Stereochem. *13*, 1 (1982); b) Heathcock, C. H.: in ref. [1 b], Vol. 3, 1984; c) Masamune, S., Choy, W.: Aldrichim. Acta *15*, 47 (1982); d) Mukaiyama, T.: Org. React. *28*, 203 (1982).
15. a) Hoffmann, R. W.: Angew. Chem. *94*, 569 (1982); Angew. Chem., Int. Ed. Engl. *21*, 555 (1982); b) Yamamoto, Y., Maruyama, K.: Heterocycles *18*, 357 (1982).
16. Heathcock, C. H., Flippin, L. A.: J. Am. Chem. Soc. *105*, 1667 (1983).
17. Jones, P. R., Goller, E. J., Kauffman, W. J.: J. Org. Chem. *36*, 3311 (1971).
18. a) Reetz, M. T., Steinbach, R., Westermann, J., Peter, R.: Angew. Chem. *92*, 1044 (1980); Angew. Chem., Int. Ed. Engl. *19*, 1011 (1980); b) Reetz, M. T., Kyung, S. H., Hüllmann, M.: Tetrahedron, in press; c) Reetz, M. T., Steinbach, R., Wenderoth, B.: Synth. Commun. *11*, 261 (1981); d) Reetz, M. T., Westermann, J.: Synth. Commun. *11*, 647 (1981).
19. Peter, R.: Diplomarbeit, Univ. Marburg 1980.
20. Reetz, M. T., Steinbach, R., Westermann, J., Peter, R., Wenderoth, B.: Chem. Ber. *118*, 1441 (1985).
21. Reetz, M. T., Steinbach, R., Westermann, J., Urz, R., Wenderoth, B., Peter, R.: Angew. Chem. *94*, 133 (1982); Angew. Chem., Int. Ed. Engl. *21*, 135 (1982); Angew. Chem. Supplement *1982*, 257.
22. Reetz, M. T., Steinbach, R., Wenderoth, B., Westermann, J.: Chem. Ind. *1981*, 541.
23. Wenderoth, B.: Dissertation, Univ. Marburg 1983.
24. Steinbach, R.: Dissertation, Univ. Marburg 1982.

25. Hoppe, D.: Angew. Chem. *96*, 930 (1984); Angew. Chem. Int. Ed. Engl. *23*, 932 (1984).
26. Reetz, M. T., Peter, R.: Tetrahedron Lett. *22*, 4691 (1981).
27. Schmidtberger, S.: Diplomarbeit, Univ. Marburg 1983.
28. Piatak, D. M., Wicha, J.: Chem. Rev. *78*, 199 (1978).
29. Makino, T., Shibata, K., Rohrer, D. C., Osawa, Y.: J. Org. Chem. *43*, 276 (1978).
30. a) Heathcock, C. H., White, C. T., Morrison, J. J., Van Derveer, D.: J. Org. Chem. *46*, 1296 (1981); b) Masamune, S., Choy, W., Kerdesky, F. A. J., Imperiali, B.: J. Am. Chem. Soc. *103*, 1566 (1981); see also lit. [62].
31. Reetz, M. T., et al.: unpublished.
32. For a review of chelation- and non-chelation-controlled additions to carbonyl compounds see, Reetz, M. T.: Angew. Chem. *96*, 542 (1984); Angew. Chem., Int. Ed. Engl. *23*, 556 (1984).
33. Still, W. C., Schneider, J. A.: Tetrahedron Lett. *21*, 1035 (1980).
34. Still, W. C., McDonald, J. H.: Tetrahedron Lett. *21*, 1031 (1980).
35. a) Lewis, M. D., Kishi, Y.: Tetrahedron Lett. *23*, 2343 (1982); b) Kishi, Y.: Pure Appl. Chem. *53*, 1163 (1981).
36. a) Inch, T. D.: Carbohydr. Res. *5*, 45 (1967); b) Hoppe, I., Schöllkopf, U.: Liebigs Ann. Chem. *1983*, 372; c) Lemieux, R. U., Wong, T. C., Thøgersen, T.: Can. J. Chem. *60*, 81 (1982); d) Fischer, J.-C., Horton, D., Weckerle, W.: Carbohydr. Res. *59*, 459 (1977); e) Paulsen, H., Roden, K., Sinnwell, V., Luger, P.: Liebigs Ann. Chem. *1981*, 2009.
37. Kelly, T. R., Kaul, P. A.: J. Org. Chem. *48*, 2775 (1983).
38. Metternich, R.: Dissertation, Univ. Marburg 1985.
39. Reetz, M. T., Kesseler, K., Schmidtberger, S., Wenderoth, B., Steinbach, R.: Angew. Chem. *95*, 1007 (1983); Angew. Chem., Int. Ed. Engl. *22*, 989 (1983); Angew. Chem. Supplement *1983*, 1511.
40. a) Kiyooka, S., Heathcock, C. H.: Tetrahedron Lett. *24*, 4765 (1983); b) Heathcock, C. H., Kiyooka, S., Blumentopf, T. A.: J. Org. Chem. *49*, 4214 (1984).
41. Reetz, M. T., Kesseler, K.: J. Chem. Soc., Chem. Commun. *1984*, 1079.
42. a) Keck, G. E., Boden, E. P.: Tetrahedron lett. *25*, 1879 (1984); b) Keck, G. E., Abbott, D. E.: Tetrahedron Lett. *25*, 1883 (1984); and lit. cited therein.
43. a) Mukaiyama, T., Banno, K., Narasaka, K.: J. Am. Chem. Soc. *96*, 7503 (1974); b) chelation-controlled addition of diketene to a chiral aldehyde: Izawa, T., Mukaiyama, T.: Chem. Lett. *1978*, 409.
44. a) Reetz, M. T., Kesseler, K., Jung, A.: Tetrahedron *40*, 4327 (1984); b) Reetz, M. T., Kesseler, K., Jung, A.: Tetrahedron Lett. *25*, 729 (1984).
45. Reetz, M. T., Jung, A.: J. Am. Chem. Soc. *105*, 4833 (1983).
46. a) Nakamura, E., Shimada, J., Horiguchi, Y., Kuwajima, I.: Tetrahedron Lett. *24*, 3341 (1983); b) Nakamura, E., Kuwajima, I.: Tetrahedron Lett. *24*, 3343 (1983).
47. Frater, G., Müller, U., Günther, W.: Tetrahedron *40*, 1269 (1984).
48. Kesseler, K.: Dissertation, Univ. Marburg 1986.
49. Reetz, M. T., Jung, A., Kesseler, K.: Angew. Chem., in press.
50. a) Danishefsky, S. J., Pearson, W. H., Harvey, D. F.: J. Am. Chem. Soc. *106*, 2456 (1984); b) Danishefsky, S. J., Pearson, W. H., Harvey, D. F., Maring, C. J., Springer, J. P.: J. Am. Chem. Soc. *107*, 1256 (1985).
51. Battioni, J. P., Chodkiewicz, W.: Bull. Soc. Chim. Fr. *1977*, 320.
52. a) Suzuki, K., Yuki, Y., Mukaiyama, T.: Chem. Lett. *1981*, 1529; b) Tabusa, F., Yamada, T., Suzuki, K., Mukaiyama, T.: Chem. Lett. *1984*, 405; and lit. cited therein.

53. a) Reetz, M. T., Kesseler, K.: J. Org. Chem., in press; see also Mead, H., MacDonald, T. L.: J. Org. Chem. *50*, 422 (1985).

54. Other types of intramolecular allyl group additions to aldehydes: a) Denmark, S. E., Weber, E. J.: J. Am. Chem. Soc. *106*, 7970 (1984); b) Mikami, K., Maeda, T., Kishi, N., Nakai, T.: Tetrahedron Lett. *25*, 5151 (1984), and lit. cited therein.

55. Jung, A.: Diplomarbeit, Univ. Marburg 1983.

56. Reetz, M. T., Jung, A., Bolm, C.: Angew. Chem., submitted.

57. Wolfram, M. L., Hanessian, S.: J. Org. Chem. *27*, 1800 (1962); although this publication lists a single diastereomer, it was later shown that actually a mixture prevails [36a].

58. Grauert, M., Schöllkopf, U.: Liebigs Ann. Chem. *1985*, 1817.

59. Mulzer, J., Angermann, A.: Tetrahedron Lett. *24*, 2843 (1983).

60. a) Fuganti, C., Servi, S., Zirotti, C.: Tetrahedron Lett. *24*, 5285 (1983); and lit. cited therein; b) Fuganti, C., Grasselli, P., Pedrocchi-Fantoni, G.: J. Org. Chem. *48*, 909 (1983).

61. Hoffmann, R. W., Endesfelder, A., Zeiss, H. J.: Carbohydrate Res. *123*, 320 (1983).

62. Masamune, S., Choy, W., Petersen, J. S., Sita, L. R.: Angew. Chem. *97*, 1 (1985); Angew. Chem., Int. Ed. Engl. *24*, 1 (1985).

63. a) Hüllmann, M.: projected Dissertation, Univ. Marburg 1986; b) Reetz, M. T., Hüllmann, M.: Tetrahedron Lett., submitted.

64. Hoffmann, R. W.: private communication.

65. Acyclic transition states have been discussed in several cases, e.g.: a) Noyori, R., Nishida, I., Sakata, J.: J. Am. Chem. Soc. *103*, 2106 (1981); b) Yamamoto, Y., Yatagai, H., Naruta, Y., Maruyama, K.: J. Am. Chem. Soc. *102*, 7107 (1980); c) Hayashi, T., Kabeta, K., Hamachi, I., Kumada, M.: Tetrahedron Lett. *24*, 2865 (1983); d) Denmark, S. E., Weber, E.: Helv. Chim. Acta *66*, 1655 (1983); and references cited in these papers.

66. a) Evans, D. A., McGee, L. R.: Tetrahedron Lett. *21*, 3975 (1980); b) Yamamoto, Y., Maruyama, K.: Tetrahedron Lett. *21*, 4607 (1980); c) Yamamoto, Y., Yatagai, H., Maruyama, K.: J. Chem. Soc., Chem. Commun. *1981*, 162; d) Harada, T., Mukaiyama, T.: Chem. Lett. *1982*, 467.

67. Peter, R.: Dissertation, Univ. Marburg 1983.

68. a) Jackman, L. M., Szeverenyi, N. M.: J. Am. Chem. Soc. *99*, 4954 (1977); b) House, H. O., Prabhu, A. V., Phillips, W. V.: J. Org. Chem. *41*, 1209 (1976); c) Vogt, H.-H., Gompper, R.: Chem. Ber. *114*, 2884 (1981).

69. Goasdoue, C., Goasdoue, N., Gaudemar, M.: Tetrahedron Lett. *25*, 537 (1984).

70. Comins, D. L., Brown, J. D.: Tetrahedron Lett. *25*, 3297 (1984).

71. Nakamura, E., Kuwajima, I.: Tetrahedron Lett. *24*, 3347 (1983).

72. Reetz, M. T.: "30 Jahre Fonds der Chemischen Industrie", Verband der Chemischen Industrie, Frankfurt, p. 29, 1980.

73. a) Dubois, J. E., Axiotis, G., Bertounesque, E.: Tetrahedron Lett. *25*, 4655 (1984); b) reactions of bis-lithium salts: Mulzer, J., Büntrup, G., Finke, J., Zippel, M.: J. Am. Chem. Soc. *101*, 7723 (1979); c) stereoselective additions of S-silyl ketene, S, N acetals: Goasdoue, C., Goasdoue, N., Gaudemar, M.: Tetrahedron Lett. *24*, 4001 (1983).

74. a) Lehnert, W.: Tetrahedron *30*, 301 (1974); b) Lehnert, W.: Tetrahedron Lett. *11*, 4723 (1970); c) Mukaiyama, T.: Pure Appl. Chem. *54*, 2456 (1982).

75. Reetz, M. T., Steinbach, R., Kesseler, K.: Angew. Chem. *94*, 872 (1982); Angew. Chem., Int. Ed. Engl. *21*, 864 (1982); Angew. Chem. Supplement *1982*, 1899.

76. Corey, E. J., Enders, D.: Tetrahedron Lett. *17*, 11 (1976); b) Corey, E. J., Enders, D.: Chem. Ber. *111*, 1337 (1978).
77. Ref. [75] erroneously lists a coupling constant of 16.9 Hz.
78. Kesseler, K.: Diplomarbeit, Univ. Marburg 1983.
79. d'Angelo, J., Pecquet-Dumas, F.: Tetrahedron Lett. *24*, 1403 (1983).
80. Schöllkopf, U.: Pure Appl. Chem. *55*, 1799 (1983).
81. Schöllkopf, U., Nozulak, J., Grauert, M.: Synthesis *1985*, 55.
82. Bardenhagen, J., Schöllkopf, U.: Liebigs Ann. Chem., in press.
83. Hoffmann, R. W., Zeiß, H.-J.: J. Org. Chem. *46*, 1309 (1981).
84. a) Widler, L., Seebach, D.: Helv. Chim. Acta *65*, 1085 (1982); b) Seebach, D., Widler, L.: Helv. Chim. Acta *65*, 1972 (1982).
85. Sato, F., Iida, K., Ijima, S., Moriya, H., Sato, M.: J. Chem. Soc., Chem. Commun. *1981*, 1140.
86. a) Sato, F., Iijima, S., Sato, M.: Tetrahedron Lett. *22*, 243 (1981); b) Kobayashi, Y., Umeyama, K., Sato, F.: J. Chem. Soc., Chem. Comman. *1984*, 621
87. Reetz, M. T., Sauerwald, M.: J. Org. Chem. *49*, 2292 (1984).
88. Yamamoto, Y., Maruyama, K.: Heterocycles *18*, 357 (1982); and lit. cited therein.
89. Yamamoto, Y., Maruyama, K.: J. Organomet. Chem. *284*, C45 (1985).
90. Reetz, M. T., Hüllmann, M., Massa, W., Berger, S., Rademacher, P., Heymanns, P.: J. Am. Chem. Soc., in press.
91. Reetz, M. T., Wenderoth, B.: Tetrahedron Lett. *23*, 5259 (1982).
92. a) Corriu, R. J. P., Lanneau, G. F., Leclercq, D., Samate, D.: J. Organomet. Chem. *144*, 155 (1978); b) Ehlinger, E., Magnus, P.: J. Am. Chem. Soc. *102*, 5004 (1980); c) Lau, P. W. K., Chan, T. H.: Tetrahedron Lett. *19*, 2383 (1978).
93. a) Tsai, D. J. S., Matteson, D. S.: Tetrahedron Lett. *22*, 2751 (1981); see also: b) Yamamoto, Y., Saito, Y., Maruyama, K.: Tetrahedron Lett. *23*, 4597 (1982); c) Yamamoto, Y., Saito, Y., Maruyama, K.: J. Chem. Soc. Chem. Commun. *1982*, 1326.
94. Sato, F., Suzuki, Y., Sato, M.: Tetrahedron Lett. *23*, 4589 (1982).
95. Ikeda, Y., Furuta, K., Meguriya, N., Ikeda, N., Yamamoto, H.: J. Am. Chem. Soc. *104*, 7663 (1982).
96. The authors in ref. [95] do not stipulate the coordination number of the reactive reagent.
97. a) Ukai, J., Ikeda, Y., Ikeda, N., Yamamoto, H.: Tetrahedron Lett. *25*, 5173 (1984); b) Ikeda, Y., Ukai, J.; Ikeda, N., Yamamoto, H.: Tetrahedron Lett. *25*, 5177 (1984).
98. Murai, A., Abiko, A., Shimada, N., Masamune, T.: Tetrahedron Lett. *25*, 4951 (1984).
99. Murai, A., Abiko, A., Masamune, T.: Tetrahedron Lett. *25*, 4955 (1984).
100. Ukai, J., Ikeda, Y., Ikeda, N., Yamamoto, H.: Tetrahedron Lett. *24*, 4029 (1983).
101. Hanko, R., Hoppe, D.: Angew. Chem. *94*, 378 (1982); Angew. Chem., Int. Ed. Engl. *21*, 372 (1982); Angew. Chem. Supplement *1982*, 961.
102. Hoppe, D., Brönneke, A.: Tetrahedron Lett. *24*, 1687 (1983).
103. v. Hülsen, E., Hoppe, D.: Tetrahedron Lett. *26*, 411 (1985).
104. Biellmann, J. F., Ducep, J. B.: Org. React. *27*, 1 (1982).
105. Roder, H., Helmchen, G., Peters, E. M., Peters, K., v. Schnering, H. G.: Angew. Chem. *96*, 895 (1984); Angew. Chem., Int. Ed. Engl. *23*, 898 (1984).

106. Patai, S. (editor): "The Chemistry of the Carbon—Carbon Triple Bond", Wiley N.Y. 1978.

107. See for example: Favre, E., Gaudemar, M.: J. Organomet. Chem. *92*, 17 (1975).

108. a) Furuta, K., Ishigura, M., Haruta, R., Ikeda, Y., Yamamoto, H.: Bull. Chem. Soc. Jap. *57*, 2768 (1984); b) Ishiguro, M., Ikeda, N., Yamamoto, H.: J. Org. Chem. *47*, 2227 (1982); for a correction see: J. Org. Chem. *48*, 142 (1983).

109. Hiraoka, H., Furuta, K., Ikeda, N., Yamamoto, H.: Bull. Chem. Soc. Jap. *57*, 2777 (1984).

110. a) Yamamoto, Y., Ito, W., Maruyama, K.: J. Chem. Soc. Chem. Commun *1984*, 1004; b) Metternich, R.: Dissertation, Univ. Marburg 1985.

111. Ashby, E. C., Laemmle, J. T.: Chem. Rev. *75*, 521 (1975).

112. Jones, P. R., Goller, E. J., Kauffman, W. J.: J. Org. Chem. *34*, 3566 (1969).

113. Houlihan, W. J.: J. Org. Chem. *27*, 3860 (1962).

114. Ashby, E. C., Noding, S. A.: J. Org. Chem. *44*, 4371 (1979).

115. Macdonald, T. L., Still, W. C.: J. Am. Chem. Soc. *97*, 5280 (1975).

116. Laemmle, J., Ashby, E. C., Roling, P. V.: J. Org. Chem. *38*, 2526 (1973).

117. Weidmann, B., Maycock, C. D., Seebach, D.: Helv. Chim. Acta *64*, 1552 (1981).

118. Gaudemar, M.: Tetrahedron *32*, 1689 (1976).

119. Abenheïm, D., Henry-Basch, E., Freon, P.: Bull. Soc. Chim. Fr. *1969*, 4038.

120. Naruta, Y., Ushida, S., Maruyama, K.: Chem. Lett. *1979*, 919.

121. Westermann, J.: Dissertation, Univ. Marburg 1982.

122. a) Mazaleyrat, J. P., Cram, D. J.: J. Am. Chem. Soc. *103*, 4585 (1981); b) Eleveld, M. B., Hogeveen, H.: Tetrahedron Lett. *25*, 5187 (1984); and lit. cited therein.

123. Weidmann, B., Widler, L., Olivero, A. G., Maycock, C. D., Seebach, D.: Helv. Chim. Acta *64*, 357 (1981).

124. Olivero, A. G., Weidmann, B., Seebach, D.: Helv. Chim. Acta *64*, 2485 (1981).

125. a) Seebach, D., Beck, A. K., Schiess, M., Widler, L., Wonnacott, A.: Pure Appl. Chem. *55*, 1807 (1983); b) Seebach, D. in: "Modern Synthetic Methods", (Scheffold, R., editor), Vol. III, Salle Verlag, Frankfurt, Verlag Sauerländer, Aarau, p. 217, 1983.

126. Sauerwald, M.: Dissertation, Univ. Marburg 1983.

127. Reetz, M. T., Kyung, S. H., Westermann, J.: Organometallics *3*, 1716 (1984).

128. Dormond, A., Tirouflet, J., le Moigne, F.: J. Organomet. Chem. *101*, 71 (1975); and lit. cited therein.

129. Reetz, M. T., Kükenhöhner, T.: unpublished.

130. Enantioselective aldol additions using chirally modified enolates: a) Evans, D. A.: in ref. [1b], Vol. 3, 1984; b) Solladie, G.: in ref. [1b], Vol. 2, 1983; c) Braun, M., Devant, R.: Tetrahedron Lett. *25*, 5031 (1984).

6. Michael Additions

Although triisopropoxy- and *tris*(dialkylamino)titanium enolates as well as alkyl- and aryltitanium reagents $RTiX_3$ (R = $OCHMe_2$, Cl) generally react with, α,β-unsaturated aldehydes and ketones in a 1,2 manner [1–3] (Chapter 3), titanium chemistry can be used to accomplish Michael additions in certain cases. A synthetically important example involves the $TiCl_4$ induced 1,4-addition of allylsilanes to α,β-unsaturated ketones (Hosomi-Sakurai reaction) [4, 5] which has been applied in numerous cases (e.g., $4 \rightarrow 5$) [4]. Since several reviews [5] have appeared, it will not be discussed here in great detail.

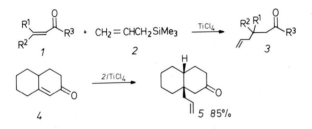

The procedure is considerably more efficient than the addition of allyl-cuprate. Mechanistically $CH_2{=}CHCH_2TiCl_3$ is probably not involved. Rather, $TiCl_4$ activates the ketone *1* by carbonyl complexation. Indeed, Lewis acids other than $TiCl_4$ are also effective. In some interesting intramolecular variations [5–7], $EtAlCl_2$ [8] has proven to be more efficient than $TiCl_4$; protons of unknown origin cause desilylation, a nuisance which can be suppressed by using "proton sponge" acids [7]. Other silylated C-nucleophiles also undergo Lewis acid mediated Michael additions, $TiCl_4$ often being the Lewis acid of choice. Much of this chemistry has been reviewed [9], and only a few examples are given here. A versatile synthesis of 1,5-diketones *7* makes use of the $TiCl_4$ induced Michael addition of enol silanes *6* to α,β-unsaturated ketones *1* [10]. Nitroolefins *8* as Michael acceptors lead to 1,4-diketones *9* (following Nef-type of work-up) [11]. It is not clear whether trichlorotitanium enolates are involved.

Enol silanes also add to α,β-ethylenic acyl cyanides *11*, a process that has been postulated to occur via trichlorotitanium enolates *10* [12]. A similar reaction with allylsilanes had previously been reported [13].

194

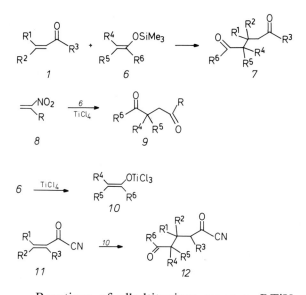

Reactions of alkyltitanium reagents $RTiX_3$ (X = Cl, $OCHMe_2$) with such Michael acceptors as α,β-unsaturated esters, ethylenic sulfones or nitro compounds have not been described. However, Michael additions of $CH_3Ti(OCHMe_2)_3$ to chiral enone sulfoxides have been studied [14]. For example, addition to the S-configurated sulfoxide *13* followed by sulfur cleavage and derivatization results in the dinitrophenylhydrazone *15* having the R-configuration (ee = 91%). Since the ee-value is higher in case of CH_3MgCl (98–100%), titanation lowers stereoselectivity [14]. Other Grignard reagents also add stereoselectively, particularly if *13* is first treated with $ZnBr_2$ [14, 15]. Chelation at the two oxygen atoms of *13* results in steric shielding of one of the diastereotopic faces [15]. Thus, it seems that titanium reagents of pronounced Lewis acidity (e.g., $RTiCl_3$) ought to perform well in such reactions (cf. Chapter 5).

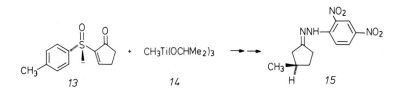

In contrast to *13*, virtually complete asymmetric induction was observed upon adding *14* to the cyclohexenone derivative *16* [15]. In this case the $CH_3MgBr/ZnBr_2$ reaction is considerably less selective (ee = 42%).

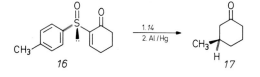

6. Michael Additions

The question of regioselective addition of carbon nucleophiles to 1-acyl-pyridimium salts has been addressed on numerous occasions [16]. For example, lithium enolates generally react with *18* to produce statistical mixtures of 1,2- and 1,4-adducts [17]. It was then discovered that the use of titanium ate complexes (made by titanating Li-enolates with Ti(OCHMe₂)₄ [18]) allows for relative good control of 1,4-addition, e.g., *18* → *20* [17]. The corresponding triisopropoxytitanium enolates are a little less regioselective. 2- and 3-picoline derivatives are also attacked at the 4-position. The adducts can be aromatized to form substituted pyridines [17].

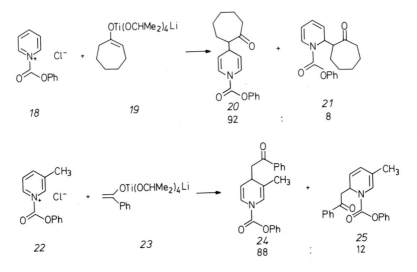

An interesting report concerning keten S-titanium-S,N-acetals in 1,2- and 1,4-additions to α,β-unsaturated ketones has appeared [19]. Whereas the tri-isopropoxytitanium enolate *27a* adds to *29* solely in a 1,2 manner, the ate complex *28a* delivers a mixture in which the Michael product is slightly favored. In case of the ate complex *28b*, clean 1,4-addition occurs (a single diastereomer!). The corresponding (Me₂CHO)₃Ti-enolate *27b* fails to add. Further experiments are necessary to understand these novel observations [19]. In case of Ti(OCHMe₂)₄ addition to *26b* there may be an equilibrium *26b* ⇌ *27b* ⇌ *28b*.

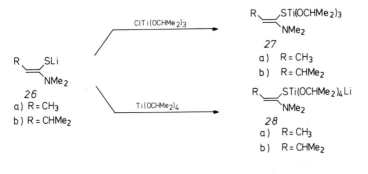

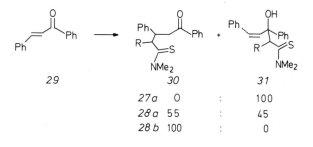

	30		31
27a	0	:	100
28a	55	:	45
28b	100	:	0

In an important study concerning a one-step $TiCl_4$ mediated cyclopentene annulation, silylallenes were added to α,β-unsaturated ketones [20]. This process involves initial Michael addition.

Finally, titanium ester enolates such as *33* are effective initiators for the controlled oligomerization of methyl methacrylate *34* [21]. At $-30\ °C$ (2 h) quantitative polymerization occurs, the molecular weight distribution being surprisingly narrow ($D = \overline{M}_w/\overline{M}_n = 1.4$). This process proceeds via iterative Michael additions and is related to the group transfer polymerization of *34* using O-silyl ketene ketals in the presence of catalysts such as tetrabutyl-ammonium or tris(dimethylamino)sulfonium fluorides [22]. The advantage of the titanium mediated polymerization has to do with the fact that no catalyst is needed. The disadvantage relates to the necessity of maintaining low temperatures. The optimum is at $-30\ °C$. At $0\ °C$ the D-value is ~ 4. The increasing value of D with increasing temperature is understandable, since chain terminating processes such as intramolecular Claisen reactions are more likely to occur under such conditions [21]. At low temperatures the polymer is living. Titanium ate complexes also polymerize *34* quantitatively (at $-78\ °C$, $D = 1.4$) [21].

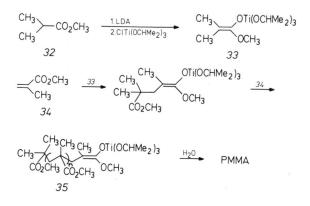

References

1. Reetz, M. T.: Top. Curr. Chem. *106*, 1 (1982).
2. Weidmann, B., Seebach, D.: Angew. Chem. *95*, 12 (1983); Angew. Chem., Int. Ed. Engl. *22*, 31 (1983).

3. A rare exception concerns tetrabenzyltitanium: Poulet, D., Casperos, J., Jacot-Guillarmod, A.: Helv. Chim. Acta *67*, 1475 (1984).

4. Hosomi, A., Sakurai, H.: J. Am. Chem. Soc. *99*, 1673 (1977).

5. Reviews: a) Sakurai, H.: Pure Appl. Chem. *54*, 1 (1982); b) Sakurai, H., Hosomi, A., Hayashi, J.: Org. Synth. *62*, 86 (1984); c) chemoselective additions: Majetich, G., Casares, A. M., Chapman, D., Behnke, M.: Tetrahedron Lett. *24*, 1909 (1983).

6. a) Wilson, S. R., Price, M. F.: J. Am. Chem. Soc. *104*, 1124 (1982); b) Majetich, G., Desmond, R., Casares, A. M.: Terahedron Lett. *24*, 1913 (1983); d) Tokoroyama, T., Tsukamoto, M., Iio, H.: Tetrahedron Lett. *25*, 5067 (1984).

7. Schinzer, D.: Angew. Chem. *96*, 292 (1984); Angew. Chem., Int. Ed. Engl. *23*, 308 (1984).

8. Trost, B. M., Coppola, B. P.: J. Am. Chem. Soc. *104*, 6879 (1982).

9. a) Weber, W. P.: "Silicon Reagents for Organic Synthesis", Springer-Verlag, Berlin 1983; b) Colvin, E.: "Silicon in Organic Synthesis", Butterworths, London 1981; c) Brownbridge, P.: Synthesis *1983*, 1 and 85.

10. Naraska, K., Soai, K., Aikawa, Y., Mukaiyama, T.: Bull. Chem. Soc. Jap. *49*, 779 (1976).

11. a) Miyashita, M., Yanami, T., Yoshikoshi, A.: J. Am. Chem. Soc. *98*, 4679 (1976); b) Miyashita, M., Yanami, T., Kumazawa, T., Yoshikoshi, A.: J. Am. Chem. Soc. *106*, 2149 (1984).

12. El-Abed, D., Jellal, A., Santelli, M.: Tetrahedron Lett. *25*, 4503 (1984).

13. El-Abed, D., Jellal, A., Santelli, M.: Tetrahedron Lett. *25*, 1463 (1984).

14. a) Posner, G. H., Mallamo, J. P., Hulce, M., Frye, L. L.: J. Am. Chem. Soc. *104*, 4180 (1982); b) Posner, G. H., Kogan, T. P., Hulce, M.: Tetrahedron Lett. *25*, 383 (1984).

15. Posner, G. H., Frye, L. L., Hulce, M.: Tetrahedron *40*, 1401 (1984).

16. See the following papers and the lit. cited therein: a) Lee, C. M., Sammes, M. P., Katritzki, A. R.: J. Chem. Soc., Perkin Trans. I *1980*, 2458; b) Akiba, K., Nishihara, Y., Wada, M.: Tetrahedron Lett. *24*, 5269 (1983); c) Commins, D. L., Smith, R. K., Stroud, E. D.: Heterocycles *22*, 339 (1984).

17. Commins, D. L., Brown, J. D.: Tetrahedron Lett. *25*, 3297 (1984).

18. a) Reetz, M. T., Peter, R.: Tetrahedron Lett. *22*, 4691 (1981); concerning related ate complexes, see: Reetz, M. T., Wenderoth, B.: Tetrahedron Lett. *23*, 5259 (1982).

19. Goasdoue, C., Goasdoue, N., Gaudemar, M.: Tetrahedron Lett. *25*, 537 (1984).

20. a) Danheiser, R. L., Carini, D. J., Basak, A.: J. Am. Chem. Soc. *103*, 1604 (1981); b) Danheiser, R. L., Carini, D. J., Fink, D. M., Basak, A.: Tetrahedron Lett. *39*, 935 (1983).

21. a) Zierke, T.: Diplomarbeit, Univ. Marburg 1984; b) Reetz, M. T., Zierke, T., in collaboration with Arlt, D., Piejko, K. E. (Bayer-A.G.).

22. Webster, O. W., Hertler, W. R., Sogah, D. Y., Farnham, W. B., Rajanbabu, T. V.: J. Am. Chem. Soc. *105*, 5706 (1983).

7. Substitution Reactions

One of the prime virtues of carbanion chemistry is the diversity of reactions possible: Grignard-type, aldol and Michael additions, oxidative dimerization as well as substitution processes at sp^3 and sp^2 hybridized C-atoms [1]. In case of the latter reactions, "resonance-stabilized" species such as ester and ketone enolates (and the nitrogen analogs), lithiated sulfones, sulfoxides, nitriles, etc. as well as hetero-atom-substituted reagents undergo smooth S_N2 reactions with primary and some secondary alkyl halides and tosylates. A synthetic gap becomes apparent upon attempting to perform these reactions with tertiary alkyl halides and certain base sensitive secondary analogs, because they are not S_N2 active. A similar situation arises in case of carbon nucleophiles lacking additional functionality. For example, $(CH_3)_2CuLi$ and higher order cuprates undergo smooth substitution reactions with primary and most secondary alkyl halides, but not with tertiary analogs. It turns out that in many cases these problems can be solved using titanium chemistry (Section 7.1). Certain titanium reagents also allow for the combination of two processes in a one-pot sequence, namely addition to carbonyl compounds followed by S_N1-type substitution of the oxygen function (Section 7.2.1). Conversely, titanium reagents are generally not nucleophilic enough to undergo S_N2-reactions with primary alkyl halides.

Substitution feactions at vinyl or aryl carbon atoms are best performed with cuprates or RMgX in combination with transition metal catalysts (Ni, Pd, Pt, etc.). Titanium reagents have not been employed in this area. An exception is the displacement of vinyl H-atoms in certain olefins (Section 7.3).

7.1 Titanium Enolates as Nucleophiles

The long-pending problem of α-*tert*-alkylation of carbonyl compounds was solved in a general way be treating the mixture of an enol silane and a *tert*-alkyl halide with $TiCl_4$ in CH_2Cl_2 [2, 3]. A comprehensive review has appeared [4].

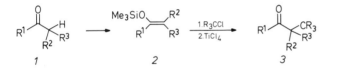

7. Substitution Reactions

is more rapid than α-*tert*-alkylation [4, 8]. For this reason, ZnX$_2$ was first introduced into this kind of chemistry [4, 9].

Secondary benzyl and substituted allyl halides (and acetates) are also S$_N$1-active and indeed can function as alkylating agents in TiCl$_4$ mediated reactions [4, 10, 11]. In a number of these cases ZnX$_2$ is superior, as originally noted for the α-*tert*-alkylation of esters [9a].

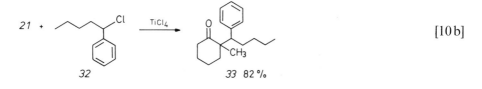

[10b]

32   33 82 %

It is clear that any S$_N$1-active alkyl halide is a potential alkylating agent in TiCl$_4$ mediated reactions [4]. These are usually exactly the compounds which are unsuitable in S$_N$2 alkylations of lithium enolates, because HX-eliminations predominate under the basic reaction conditions. Thus, the methods are complementary [4]. Indeed, S$_N$1-inactive halides such as CH$_3$I or even (CH$_3$)$_2$CHBr do not participate in TiCl$_4$ promoted α-alkylations. A final example refers to anchimerically accellerated alkylations [4] such as *34 → 35* which occur with complete retention of configuration [12]. Crossed aldol type additions using acetals *36 → 37* had previously been reported and may also involve carbocations [13] (see also Section 7.2.3).

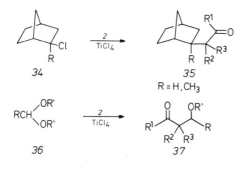

34   35
                            R = H, CH$_3$

36   37

How do the less Lewis acidic triisopropoxy- and *tris*(dialkyl)amino-titanium enolates [14] behave in reactions with primary, secondary and tertiary alkyl halides? To this end, *38* was reacted under various conditions with CH$_3$I, CH$_2$ = CHCH$_2$Br and (CH$_3$)$_3$CCl [15]. Under no circumstances could C—C bond formation be induced. The same applies to attempted reactions with epoxides and prenyl acetate in the presence of Pd(PPh$_3$)$_4$, although ligands at titanium have not been varied [16]. In contrast, the more reactive acid chlorides rapidly form 1,3-diketones, e.g., *38 → 43* [15]; however, classical acylation methods are superior [17]. Tris(dialkylamino)tita-nium enolates also fail to provide α-alkylated products in reactions with S$_N$2-active alkyl halides [15].

202

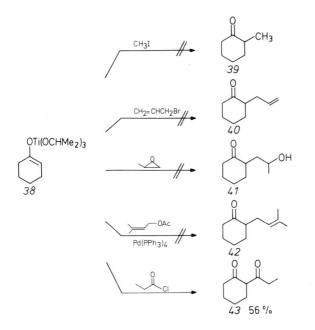

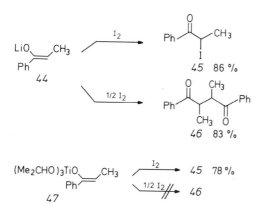

The result show that nucleophilicity of triisopropoxy- and tris(dialkyl-amino)titanium enolates is insufficient for S$_N$2 reactions [15]. This again emphasizes the complementary nature of reactions of lithium enolates with S$_N$2 alkylating agents on the one hand, and TiCl$_4$ mediated C—C bond formation using enol silanes and S$_N$1-active alkylating agents on the other [4]. This also becomes apparent in the I$_2$-induced coupling of enolates. Both the lithium and the titanium enolate (44 and 47, respectively) react with I$_2$ to provide excellent yields of the α-iodo ketone 45. However, upon using half of one equivalent of I$_2$, the lithium enolate couples to form 46 (via nucleophilic reaction between 44 and 45) [18], while the titanium enolate does not undergo oxidative dimerization [15].

These interesting C—C bond forming reactions are analogous to Lewis acid promoted alkylations of (arene)tricarbonylchromium complexes using enol silanes, which are also 100% stereoselective [41]. Me₃SiCN undergoes the same type of reaction [36]. Chiral carbocations are involved [41].

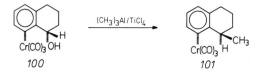

$(CH_3)_3Al / TiCl_4$

100   101

7.2.2 Direct Geminal Dialkylation of Ketones and Aldehydes and Exhaustive Methylation of Acid Chlorides

The geminal dimethyl structural unit occurs frequently in terpenes, steroids and compounds of theoretical interest. A number of strategies based on multistep procedures have been described which allow for the construction of quaternary carbon atoms [27]. Geminal dimethylation (or dialkylation) of ketones 66 → 102 is an attractive approach. It can be realized by the three step sequence as described in Section 7.2.1 or by Wittig methylenation (Chapter 8) followed by Simmons-Smith reaction and hydrogenolysis of the cyclopropane ring [27]. This section describes the direct geminal dimethylation 66 → 102 using $(CH_3)_2TiCl_2$ (55).

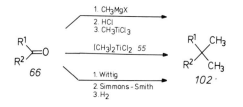

Since the Ti—O bond is very strong (Chapter 2), any reaction which forms this bond should have a pronounced thermodynamic driving force. The position specific replacement of oxygen in ketones by two methyl groups might be expected to be possible using methyltitanium reagents, because two new Ti—O bonds would be formed. Upon testing various Ti-reagents, it was discovered that ketones react with $(CH_3)_2TiCl_2$ (55) in CH_2Cl_2 to form high yields of the geminal dimethylated products 102 [37]. The mechanism involves addition (Chapter 3) followed by S_N1 ionization of 103 and capture of the intermediate carbocation 104 by non-basic methyltitanium species [42]. Optimum yields are obtained by using two parts of 55 per part of ketone. It may be recalled that species 103 are identical to those proposed in the direct methylation of alcohols using 55 (Section 7.2.1).

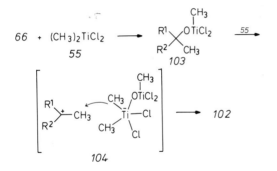

$$66 + (CH_3)_2TiCl_2 \xrightarrow{\quad} \underset{103}{R^2 \diagdown} \quad \xrightarrow{55}$$

55

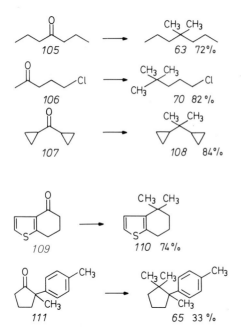

A variety of ketones can be geminal dimethylated by this procedure ($-30\ °C \rightarrow 0\ °C$), some of which are shown below [37, 42].

This one-pot procedure is thus simpler than the three step sequence described in Section 7.2.1. In fact, the tertiary chloride is sometimes too sensitive for preparation, e.g., in case of *109* the three step method failed [23c]. The geminal dimethylation of *111* constitutes a very simple synthesis of ($\pm$)-cuparene [37, 42]. The plausible assumption that carbocations are involved was recently supported by the use of optically active ketone *111*; the product turned out to be racemic, because rapid Wagner-Meerwein rearrangement of the intermediate cation causes racemization [43]. Another piece of mechanistic evidence for the intermediacy of carbocations (which also illustrates the limitation of the method) relates to the reaction of α,β-

211

Some typical examples are shown below [63]. Amines *171–173* originate from the corresponding aromatic aldehydes and methyllithium. The reaction of the Li-enolates proceeds with appreciable diastereoselectivity in relevant cases. For example, *176* is formed as a 92:8 diastereomer mixture (the relative configuration has not been established). These intriguing reactions are the first diastereoselective syntheses of Mannich bases known in the literature. Although mechanistic details need to be studied, the intermediacy of iminium salts $R_2\overset{+}{N}=CHR$ is likely [63]. Thus, the substitution reactions bear some mechanistic resemblance to the methylation of tertiary alkoxides *128* by $(CH_3)_2TiCl_2$ (Section 7.2.2).

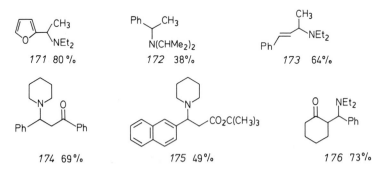

171 80 %	*172* 38%	*173* 64%

174 69%	*175* 49%	*176* 73%

Mechanistically related (but of limited synthetic value) are DIBAL-induced reductions of titanated N,O-hemiketals *178* [65], which also proceed via iminium ions [15, 36].

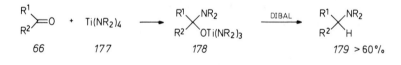

66	*177*	*178*	*179* > 60%

7.3 Other Substitution Reactions: Present and Future

It is clear that most substitution reactions using titanium reagents occur only under Lewis acidic conditions, S_N1-processes being involved. In case of acetals, complexation of the Lewis acidic titanium reagent at the alkoxy group may trigger an S_N2-type of displacement, although intimate ion pairs may be short-lived intermediates. It is likely that more substitution reactions under Lewis acidic conditions will be discovered in the future. Potentially, exchange of lithium in carbanionic species by $TiCl_3$ should lead to species which react in such a way [5c]. Studies relating to epoxides have recently been initiated [66]. Preliminary experiments with *180* show that allyltitanium compounds react cleanly at the α-position [66, 67]. Since allylmagnesium chloride delivers a mixture of regioisomers *181* and *182*, titanation increases selectivity. Diallylzinc also forms *181* [68]. Alkyl-substituted epoxides appear not to react as cleanly. In such cases it could be of interest to direct substitution at the more highly substituted C-atom, since cuprates show the opposite regioselectivity [19].

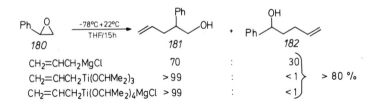

180		**181**	**182**
CH₂=CHCH₂MgCl		70 :	30
CH₂=CHCH₂Ti(OCHMe₂)₃	>99 :	<1	> 80 %
CH₂=CHCH₂Ti(OCHMe₂)₄MgCl	>99 :	<1	

S$_N$2-Displacements of the type known in classical carbanion chemistry using RX have not been observed to date, despite several attempts (Section 7.2.2). Triisopropoxy- and tris(diethylamino)titanium reagents (e.g., enolates) are not nucleophilic enough. Perhaps the use of ligands such as Cp groups at titanium could render the reagents more nucleophilic, e.g., bis(cyclo-pentadienyl)titanium enolates. Cp ligands are known to be strongly electron-releasing (Chapter 2). However, there would have to be some advantage relative to lithium reagents. Another way to increase nucleophilicity would be to deprotonate titanium reagents to form mixed di-metallic species. Initial experiments involving lithiation of titanated sulfones appear promising [69].

Another area which may turn out to be of synthetic interest concerns displacement of vinyl and aryl hydrogen atoms (Heck-type [70] of substitutions). A few cases are known, e.g., 183 → 184 [71] and 185 → 186 [72] (a pheromone). The mechanism of such displacements is unclear, but may be related to the Ziegler-Natta polymerization [71] and/or Cp₂TiCl₂ or TiCl₄ mediated carbo- and hydrometallations of olefins and acetylenes [73] (Chapter 1).

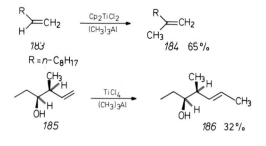

In fact, the formation of 186 is based on previous observations concerning the reaction of homo-allylic alcohols with R₃Al in the presence of TiCl₄ or other titanium compounds [73]: Addition and vinyl-H substitutions occur, the latter via an addition/β-hydride elimination mechanism. It would be of synthetic interest to control such processes in such a way that only regioselective substitution of vinyl-hydrogen atoms occurs.

50. a) Casara, P., Metcalf, B. W.: Tetrahedron Lett. *19*, 1581 (1978); b) Schmid, R., Huermann, P. L., Johnson, W. S.: J. Am. Chem. Soc. *102*, 5122 (1980).
51. a) Utimoto, K., Wakabayashi, Y., Horiie, T., Inoue, M., Shishiyama, Y., Obayashi, M., Nozaki, H.: Tetrahedron *39*, 967 (1983); b) Reetz, M. T., Chatziiosifidis, I., Künzer, H., Müller-Starke, H.: Tetrahedron *39*, 961 (1983).
52. McNamara, J. M., Kishi, Y.: J. Am. Chem. Soc. *104*, 7371 (1982).
53. Bartlett, P. A., Johnson, W. S., Elliott, J. D.: J. Am. Chem. Soc. *105*, 2088 (1983).
54. a) Hoffmann, R. W., Herold, T.: Chem. Ber. *114*, 375 (1981); b) Brown, H. C., Jadhav, P. K.: J. Org. Chem. *49*, 4089 (1984); and lit. cited therein.
55. Choi, V. M. F., Elliot, J. D., Johnson, W. S.: Tetrahedron Lett. *25*, 591 (1984).
56. Johnson, W. S., Crackett, P. H., Elliott, J. D., Jagodzinski, J. J., Lindell, S. D., Natarajan, S.: Tetrahedron Lett. *25*, 3951 (1984).
57. Johnson, W. S., Elliott, R., Elliott, J. D.: J. Am. Chem. Soc. *105*, 2904 (1983).
58. Lindell, S. D., Elliott, J. D., Johnson, W. S.: Tetrahedron Lett. *25*, 3947 (1984).
59. See for example: a) Mukaiyama, T., Soai, K., Sato, T., Shimizu, H., Suzuki, K.: J. Am. Chem. Soc. *101*, 1455 (1979); b) Mazaleyrat, J. P., Cram, D. J.: J. Am. Chem. Soc. *103*, 4585 (1981); c) Eleveld, M. B., Hogeveen, H.: Tetrahedron Lett. *25*, 5187 (1984); d) Meyers, A. I., Harre, M., Garland, R.: J. Am. Chem. Soc. *106*, 1146 (1984); e) Mori, A., Fujiwara, J., Maruoka, K., Yamamoto, H.: Tetrahedron Lett. *24*, 4581 (1983); f) Midland, M. M., Kazubski, A.: J. Org. Chem. *47*, 2495 (1982); g) Noyori, R., Tomino, I., Tanimoto, Y.: J. Am. Chem. Soc. *101*, 3129 (1979).
60. Mashraqui, S. H., Kellogg, R. M.: J. Org. Chem. *49*, 2513 (1984).
61. Mori, A., Maruoka, K., Yamamoto, H.: Tetrahedron Lett. *25*, 4421 (1984).
62. Ghribi, A., Alexakis, A., Normant, J. F.: Tetrahedron Lett. *25*, 3082 (1984).
63. a) Seebach, D., Beck, A. K., Schiess, M., Widler, L., Wonnacot, A.: Pure Appl. Chem. *55*, 1807 (1983); b) Seebach, D., Betschart, C., Schiess, M.: Helv. Chim. Acta *67*, 1593 (1984).
64. a) Comins, D. L., Brown, J. D.: Tetrahedron Lett. *22*, 4213 (1981); b) Comins, D. L., Brown, J. D., Mantlo, N. B.: Tetrahedron Lett. *23*, 3979 (1982).
65. Reetz, M. T., Wenderoth, B., Peter, R.: J. Chem. Soc., Chem. Commun. *1983*, 406.
66. Wenderoth, B.: Dissertation, Univ. Marburg 1983.
67. Reetz, M. T.: Top. Curr. Chem. *106*, 1 (1982).
68. a) Abenhaim, D., Henry-Basch, E., Freon, P.: Bull. Soc. Chim. Fr. *1970*, 179; b) Courtois, G., Miginiac, L.: J. Organomet. Chem. *69*, 1 (1974).
69. If carbonyl compounds are added, olefination sets in: M. T. Reetz, et al., unpublished.
70. a) Plevyak, J. E., Heck, R. F.: J. Org. Chem. *43*, 2454 (1978); b) Heck, R. F.: Pure Appl. Chem. *50*, 691 (1978).
71. Barber, J. J., Willis, C., Whitesides, G. M.: J. Org. Chem. *44*, 3603 (1979).
72. Schlosser, M., Fujita, K.: Angew. Chem. *94*, 320 (1982); Angew. Chem., Int. Ed. Engl. *21*, 309 (1982); Angew. Chem. Supplement *1982*, 646.
73. a) Youngblood, A. V., Nichols, S. A., Coleman, R. A., Thompson, D. W.: J. Organomet. Chem. *146*, 221 (1978); b) Schultz, F. W., Ferguson, G. S., Thompson, D. W.: J. Org. Chem. *49*, 1736 (1984); c) Brown, D. C., Nichols, S. A., Gilpin, A. B., Thompson, D. W.: J. Org. Chem. *44*, 3457 (1979); d) Sato, F., Tomuro, Y., Ishikawa, H., Sato, M.: Chem. Lett. *1980*, 99; e) Negishi, E.: Pure Appl. Chem. *53*, 2333 (1981); f) Hayami, H., Oshima, K., Nozaki, H.: Tetrahedron Lett. *25*, 4433 (1984); g) Richey, H. G., Jr., Moses, L. M., Hangeland, J. J.: Organometallics *2*, 1545 (1983) and lit. cited therein.

8. Wittig-type Methylenation of Carbonyl Compounds

The success of the Wittig reaction of phosphorus ylides with carbonyl compounds is undisputed [1]. However, in case of enolizable ketones, the yields are often poor, or epimerization of chiral centers occurs under the basic conditions. Furthermore, chemoselective olefination of ketones containing additional sensitive functionality is not always possible. In case of methylenation, such problems can be solved using powerful new methods based on titanium chemistry.

The addition of a methylene chloride solution of $TiCl_4$ (0.7 parts) to a mixture of CH_2Br_2 (1 part) and zinc dust (3 parts) in THF at room temperature leads within 15 minutes to a reagent which smoothly olefinates a variety of ketones [2]. The yields are often better than in case of the classical Wittig olefination using $Ph_3P=CH_2$:

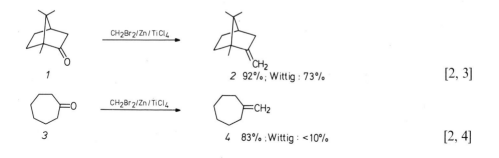

[2, 3]

[2, 4]

The procedure has been applied successfully in natural products chemistry. For example, conversion of 5 into 6 (a key intermediate in the synthesis of iridoid mono-terpenes) is almost quantitative; the classical Wittig reaction proceeds poorly due to the additional functionality present in the molecule [5].

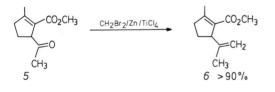

223

8. Wittig-type Methylenation of Carbonyl Compounds

Even more impressive is the conversion *7 → 8*, which fails completely using $Ph_3P=CH_2$ [6].

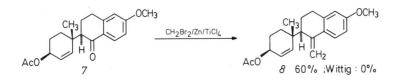

Upon attempting to apply the reaction to the synthesis of gibberellins, difficulties were encountered [7]. However, a variation of the original procedure turned out to be highly efficient [7]. The reagent is prepared from CH_2Br_2 (1 part), zinc dust (3 parts) and $TiCl_4$ (0.73 parts) in THF at low temperatures and is then allowed to age at 5 °C for three days. The resultant gray slurry reacts smoothly with ketones, including those which are enolizable and/or which contain sensitive functionality. For example, conversion of the nor-gibberellin ketone *9* into *10* is almost quantitative [7]. The yields in case of aldehydes are lower due to competing pinacol formation.

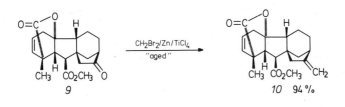

For methylenation of other nor-gibberellin ketones, e.g., *11*, it had previously been found essential to protect the 3-hydroxy function so as to avoid epimerization [8]. This is not necessary in case of the procedure based on $CH_2Br_2/Zn/TiCl_4$ (aged):

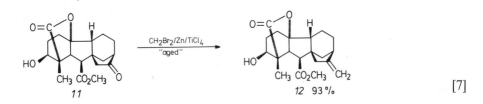

[7]

Specific labelling with deuterium using $CD_2Br_2/Zn/TiCl_4$ is also possible, a process that avoids scrambling [7]. In another application, (+)-isomenthone (*13*) was converted into (+)-3-methylene-*cis*-p-menthane (*14*) without epimerization [9]. Further examples relate to prostaglandin syntheses, e.g., *15 → 16* [10].

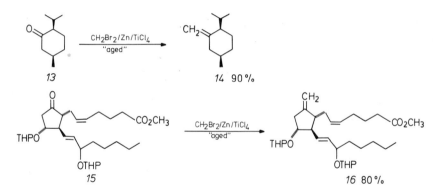

13   14 90%

15   16 80%

Presently, it is difficult to evaluate the role of reagent-aging which has been claimed to increase the efficiency of methylenation [7], because no systematic comparative studies have been reported. It is also not clear whether the specified time of aging (3 days) can be reduced. In the original and in the improved procedure, the ratio of components is chosen so that a species formally equivalent to *17* results. However, in view of the ready titanation or organylzinc reagents by $TiCl_4$ (e.g., $(CH_3)_2Zn + TiCl_4 \rightarrow (CH_3)_2TiCl_2$ [11]), an oligomeric species *18* should also be considered. In any case, the non-basic Lewis acidic reagent is likely to exist in the form of octahedral THF-adducts (cf. Chapter 2). More work is necessary to clear up the synthetic, structural and mechanistic aspects. It would also be of synthetic interest to intercept the initial ketone adduct (prior to olefin-forming fragmentation) with electrophiles such as I_2, aldehydes or H_2O. However, initial experiments show that fragmentation is at least as fast as addition [12].

$$Zn(CH_2TiCl_3)_2$$

17

$$\overset{\overset{\displaystyle Cl}{|}}{\underset{\underset{\displaystyle Cl}{|}}{\text{---CH}_2\overset{}{Ti}}}\overset{\overset{\displaystyle Cl}{|}}{\underset{\underset{\displaystyle Cl}{|}}{\text{CH}_2\overset{}{Ti}}}\overset{\overset{\displaystyle Cl}{|}}{\underset{\underset{\displaystyle Cl}{|}}{\text{CH}_2\overset{}{Ti}\text{---}}}$$

18

The so-called Tebbe reagent *21* [13] is an isolable, well characterized compound which also olefinates carbonyl compounds, including those which are highly enolizable [14, 15]. For example, the yield in case of β-tetralone (*22*) is 84% [15].

$$Cp_2TiCl_2 \quad + \quad (CH_3)_3Al \quad \longrightarrow \quad Cp_2Ti\underset{\displaystyle Cl}{\overset{\displaystyle CH_2}{<}}\underset{\displaystyle CH_3}{\overset{\displaystyle CH_3}{Al}}$$

19   20   21

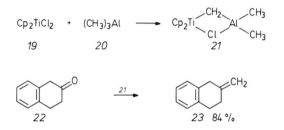

22   23 84%

225

Another virtue of *21* as a methylenating reagent pertains to the reaction with esters and lactones leading to vinyl ethers in excellent yields [16]. The reaction is of considerable synthetic interest because direct methylenation of esters using phosphorus ylides does not constitute a generally viable synthetic method. The examples shown below document the generality of this novel process [16]. Olefin positional and stereochemical integrity is maintained. Keto-esters can be bis-methylenated (*30 → 31*) [16]. Chemoselective olefination at the keto group is possible by using only one equivalent of reagent [14].

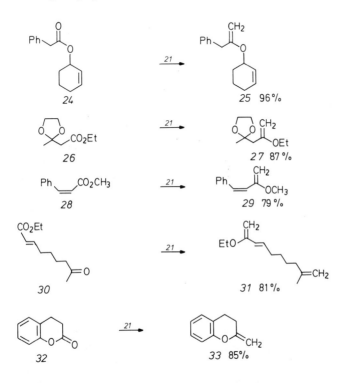

Although related transition metal complexes (e.g., of tantalum) also undergo some of these reactions [17], the titanium based procedure appears to be best [14–16]. It synthetic potential is obvious. For example, products of the type *25* are suited for Claisen rearrangements, which make possible far reaching molecular changes in two simple steps [14]. Intramolecular Diels-Alder reaction can also be envisioned for trienes of the kind *31*. Such strategies have in fact been described [16, 18a, c]. Application of *21* as a methylenating agent in carbohydrate chemistry has also been described [18b].

Related methylenating reagents such as *35* [19] and *37* [14, 15, 20, 21], have been prepared and their application in organic synthesis tested. A convenient and very mild source of the reactive intermediate $Cp_2Ti=CH_2$ (*38*) is the titanacyclobutane *37a*. It is not as Lewis acidic as *19*, and can thus be used in case of acid sensitive substrates [14] (e.g., *19* polymerizes valerolactone). *37a* and *37b* methylenate ketones under very mild conditions [15, 21]. It remains to be seen which of the various titanium based methylenating agents will be used most. One of the main advantages of $CH_2Br_2/Zn/TiCl_4$ is its ready availability.

$$CH_2(ZnI)_2 \xrightarrow{\;19\;} Cp_2TiCH_2ZnX_2$$

$$\quad\; 34 \qquad\qquad\qquad 35$$

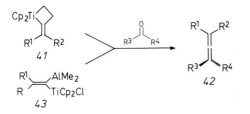

$$\begin{array}{ccc} 36 & 37 & 38 \end{array}$$

a R = CH₃
b R = n-C₃H₇

Titanacyclobutanes (e.g., *37*) form a fascinating chapter in titanium chemistry [14, 20–22]. Besides their role in olefin metathesis [14, 20], they methylenate acid chlorides, e.g., *39* → *40* [14, 23]; a similar reaction occurs in case of *19* [24]. The resulting bis(cyclopentadienyl)titanium enolates can be used in a variety of ways, e.g., aldol additions [14, 23] or alkylation with CH_3I (!) [24] (cf. Chapter 7 concerning substitution reactions of titanium enolates).

$$\underset{39}{\overset{O}{\underset{\parallel}{RCCl}}} \xrightarrow{\;37\;} \underset{40}{CH_2{=}\!\!<\!\!\begin{array}{l}OTiCp_2Cl\\ R\end{array}}$$

Several derivatives of *19* and *37* have been described [14]. For example, *41* reacts with ketones to afford allenes *42* under mild conditions [25a]. Previously, *43* had been shown to perform well as alkylidene transfer reagents [25b]. Certain zirconium-based 1,1-dimetalloalkanes react with ketones to form olefins [26].

A number of titanium mediated Peterson olefinations have been reported. Reagent *45* adds chemoselectively to aldehydes in the presence of ketones, leading to moderate yields of terminal olefins *46* [27]. The reaction

227

requires a three-fold excess of reagent for optimum yields. The less Lewis acidic analog *47* is considerably less reactive [28, 29]. It is also aldehyde-selective (Chapter 3), but does not induce olefination under the reaction conditions used. However, the adducts *48* are easily converted into *46* under basic or acidic conditions [29]. Methylenation of aldehydes are perhaps best carried out using the classical Wittig reaction [1] or molybdenum reagents [30]; the latter are effective in protic solvents!

$$Me_3SiCH_2MgCl \xrightarrow[Et_2O]{TiCl_4} Me_3SiCH_2TiCl_3 \xrightarrow{RCHO} RCH=CH_2$$

$$\textit{44} \qquad\qquad\qquad \textit{45} \qquad\qquad\qquad \textit{46}$$

$$Me_3SiCH_2Ti(OCHMe_2)_3 \xrightarrow[2)\,H_2O]{1)\,RCHO} \overset{OH}{\underset{|}{R}}CHCH_2SiMe_3$$

$$\textit{47} \qquad\qquad\qquad\qquad \textit{48}$$

A number of trimethylsilyl-substituted allyllithium reagents have been titanated. The resulting species react regio- and stereoselectively to provide dienes (Chapter 5) 1,1-Dimetallo reagents involving Ti and Al are also known [31].

References

1. a) Maercker, A.: Org. React. *14*, 270 (1965); b) Johnson, A. W.: "Ylid Chemistry", Academic Press, N.Y. 1966.
2. a) Takai, K., Hotta, Y., Oshima, K., Nozaki, H.: Tetrahedron Lett. *19*, 2417 (1978); b) Takai, K., Hotta, Y., Oshima, K., Nozaki, H.: Bull. Chem. Soc. Jap. *53*, 1698 (1980).
3. Greenwald, R., Chaykovsky, M., Corey, E. J.: J. Org. Chem. *28*, 1128 (1963).
4. Schriesheim, A., Muller, R. J., Rowe, C. A., jr.: J. Am. Chem. Soc. *84*, 3164 (1962).
5. Imagawa, T., Sonobe, T., Ishiwara, H., Akiyama, T.: J. Org. Chem. *45*, 2005 (1980).
6. Mincione, E., Pearson, A. J., Bovicelli, P., Chandler, M., Heywood, G. C.: Tetrahedron Lett. *22*, 2929 (1981).
7. Lombardo, L.: Tetrahedron Lett. *23*, 4293 (1982).
8. Lombardo, L., Mander, L. N., Turner, J. V.: J. Am. Chem. Soc. *102*, 6626 (1980).
9. Lombardo, L., submitted to Org. Synth.
10. a) Shibasaki, M., Torisawa, Y., Ikegami, S.: Tetrahedron Lett. *24*, 3493 (1983); b) Ogawa, Y., Shibasaki, M.: Tetrahedron Lett. *25*, 1067 (1984).
11. Reetz, M. T., Westermann, J., Steinbach, R.: Angew. Chem. *92*, 931 (1980); Angew. Chem., Int. Ed. Engl. *19*, 900 (1980).
12. Reetz, M. T., von Idzstein, M., unpublished.
13. Tebbe, F. N., Parshall, G. W., Reddy, G. S.: J. Am. Chem. Soc. *100*, 3611 (1978).

14. Review of the use of the Tebbe reagent and other sources of $Cp_2Ti=CH_2$: Brown-Wensley, K. A., Buchwald, S. L., Cannizzo, L., Clawson, L., Ho, S., Meinhardt, D., Stille, J. R., Straus, D., Grubbs, R. H.: Pure Appl. Chem. *55*, 1733 (1983).
15. Clawson, L., Buchwald, S. L., Grubbs, R. H.: Tetrahedron Lett. *25*, 5733 (1984).
16. Pine, S. H., Zahler, R., Evans, D. A., Grubbs, R. H.: J. Am. Chem. Soc. *102*, 3272 (1980).
17. Schrock, R. R.: J. Am. Chem. Soc. *98*, 5399 (1976).
18. a) Ireland, R. E., Varney, M. D.: J. Org. Chem. *48*, 1829 (1983); b) Wilcox, C. S., Long, G. W., Suh, H.: Tetrahedron Lett. *25*, 395 (1984). c) Kinney, W. A., Coghlan, M. J., Paquette, L. A.: J. Am. Chem. Soc. *106*, 6868 (1984).
19. Eisch, J. J., Piotrowski, A.: Tetrahedron Lett. *24*, 2043 (1983).
20. Lee, J. B., Gojda, G. J., Schaefer, W. P., Howard, T. R., Ikariya, T., Straus, D. A., Grubbs, R. H.: J. Am. Chem. Soc. *103*, 7358 (1981).
21. Seetz, W. F. L., Schat, G., Akkerman, O. S., Bickelhaupt, F.: Angew. Chem. *95*, 242 (1983); Angew. Chem, Int. Ed. Engl. *22*, 248 (1983); Angew. Chem. Supplement *1983*, 234.
22. Mackenzie, P. B., Ott, K. C., Grubbs, R. H.: Pure Appl. Chem. *56*, 59 (1984).
23. Stille, J. R., Grubbs, R. H.: J. Am. Chem. Soc. *105*, 1664 (1983).
24. Chou, T. S., Huang, S. B.: Tetrahedron Lett. *24*, 2169 (1983).
25. a) Buchwald, S. L., Grubbs, R. H.: J. Am. Chem. Soc. *105*, 5490 (1983); b) Yoshida, T., Negishi, E.: J. Am. Chem. Soc. *103*, 1276 (1981).
26. Hartner, F. W., jr., Schwartz, J.: J. Am. Chem. Soc. *103*, 4979 (1981).
27. Kauffmann, T., König, R., Pahde, C., Tannert, A.: Tetrahedron Lett. *22*, 5031 (1981).
28. Reetz, M. T.: Top. Curr. Chem. *106*, 1 (1982).
29. Reetz, M. T., Westermann, J., Steinbach, R., Wenderoth, B., Peter, R., Ostarek, R., Maus, S.: Chem. Ber. *118*, 1421 (1985).
30. a) Kauffmann, T., Kieper, G.: Angew. Chem. *96*, 502 (1984); Angew. Chem., Int. Ed. Engl. *23*, 532 (1984); b) Kauffmann, T., Fiegenbaum, P., Wieschollek, R.: Angew. Chem. *96*, 500 (1984); Angew. Chem., Int. Ed. Engl. *23*, 531 (1984).
31. Yoshida, T.: Chem. Lett. *1982*, 429, and lit. cited therein.

Subject Index

Subject Index

Transesterification, using Ti(OR)$_4$
 10–11
Trimers   45–46
Trisdialkylaminoalkyltitanium
 compounds   26–27
α, β-Unsaturated carbonyl compounds,
 1,2-vs. 1,4-addition to   1, 7, 88–90,
 103, 162, 194–197
β, γ-Unsaturated carbonyl compounds,
 synthesis of   91

UV spectra   50–51, 56, 60

Vinyl ethers, synthesis of   226

Wittig-like olefinations   1, 3, 35, 82,
 102, 210, 214, 223–228
Workup procedures   100
Wagner-Meerwein
 rearrangements   207–208, 211–212

Ziegler-Natta polymerization   3–4,
 19–20, 22, 51, 57, 219
Zirconium enolates   150, 153